DIGITAL TECHNIQUES AND MICROPROCESSOR SYSTEMS

Digital Techniques and Microprocessor Systems

Servicing Electronic Systems Volume 3

*A Textbook for the City and Guilds of London Institute
Course Nos 2240 & 7261 and for BTEC U86/343 & 347*

IAN R. SINCLAIR BSc
*Formerly Lecturer in Physics & Electronics
Braintree (Essex) College of Further Education*

GEOFFREY E. LEWIS MIEEE, BA, MSc, MRTS, MIEEIE
*Formerly Senior Lecturer in Radio, Television and Electronics,
Canterbury College of Technology*

Avebury Technical

Aldershot • Brookfield USA • Hong Kong • Singapore • Sydney

Published by
Avebury
Ashgate Publishing Limited
Gower House
Croft Road
Aldershot
Hants GU11 3HR
England

Ashgate Publishing Company
Old Post Road
Brookfield
Vermont 05036
USA

British Library Cataloguing in Publication Data

Sinclair, Ian R.
    Digital Techniques and Microprocessor Systems.
    Servicing Electronic Systems. – Vol. 3
    I. Title II. Lewis, Geoffrey E.
    621.3810288

    ISBN 0 291 39834 0

Library of Congress Catalog Card Number: 95-80358

Printed in Great Britain at the University Press, Cambridge

Contents

Preface

Whilst this book follows the successful format of the earlier volumes, it marks a departure from the usual structure of the Servicing Electronic Systems series. Due to the large amount of common core material it has been possible to cater for Digital Techniques and Microprocessor Computer Systems, a popular combination for Part 3 of C&G 2240 in one book. Furthermore, we have also been able to provide a very significant coverage for the BTEC courses, Data Communications NIII and Information Handling Systems N, together with the C&G 7261 Information Technology Scheme, Microcomputer Systems, Part III. This has been achieved by the periodic reference back to earlier volumes for the revision of certain topics to avoid a significant amount of text duplication. Because of the wide coverage, this volume has a significant hardware and software content.

The codes that have been used at the head of each chapter to identify the various syllabus sections are as follows:

B1 = BTEC Module U86/343 Information Handling Systems N
B2 = BTEC Module U86/347 Data Communications NIII
D = Digital Techniques, Part 3, C&G 2240
I = Information Technology Scheme, Microcomputer Systems Installation and Maintenance III, C&G 7261
M = Microprocessor Computer Systems, Part 3, C&G 2240

For example, D01–2.2 refers to Digital Techniques, Section 01, paragraph 2.2

Many of today's established systems have been concentrated onto applications specific integrated circuits (ASICs). This gives the advantages of improved reliability, reduced system cost and size, which in turn leads to a greater popularity of the systems. In spite of these developments, discrete component circuits have been retained in many cases because it is felt that with these it is easier to explain the principles of operation.

The traditional boundaries between communications, information processing and computing systems have virtually disappeared; the term communications is thus now an all-embracing one. The reason for this convergence is chiefly due to the recent developments in semiconductor and electronic technology, together with the economic availability of powerful computing systems. This in turn makes the work of the service technician more complex but at the same time, more rewarding.

The approach used in this book is again essentially practical with the minimum of mathematics, placing the emphasis on experiment and faultfinding. The authors have been mindful not only of the special needs of the home based or distance learner but also of the need to provide the important underpinning knowledge that is demanded by the certification of vocational qualifications. The text can therefore be used as a self study course for students and readers who wish to extend their knowledge of computer based systems in general. In particular, the chapter on PC machines should prove to be valuable to the many users of these very popular systems.

A number of computer programs that are designed to exercise either peripheral devices or I/0 interfaces have been included. Some of these owe their origins to either Ken Taylor, Mike Tooley or Ron Vears, to whom the authors would like to express their appreciation.

A guide book has been prepared which contains useful course hints together with comments on the questions and exercises included in the main text. This booklet, which may be freely photocopied, is available free of charge to lecturers and instructors from:

Customer Services,
Avebury Technical,
Gower Publishing Co. Ltd,
Gower House,
Croft Road,
Aldershot,
Hants. GU11 3HR.

1   Chip technology

Syllabus references: *D01, I1–1.1 & 1.2*

Gate inputs and outputs

NOTE: *Before* reading this chapter, you should revise Chapter 9 of Volume 2, Part 1, of this series, particularly the sections dealing with gates and gate types.

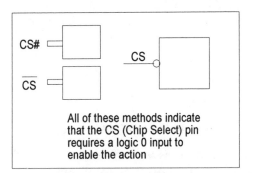

All of these methods indicate that the CS (Chip Select) pin requires a logic 0 input to enable the action

Figure 1.1   Methods of indicating that an active signal is at logic 0 level

Throughout this book, several methods of indicating a logic low input or output will be used, as shown here. All of these methods are likely to be found in

1

manufacturers' literature and in examination papers, and you should know all three. The most recent is the use of the # (hash) sign, and this is likely to become established because it can be typed and printed easily.

Digital integrated circuits (ICs) operate using dc supplies that are always voltage-stabilised, and which usually also require local decoupling, such as a 0.1μF capacitor connected from the supply voltage pin of the IC to an earth line. In critical circuits these decoupling capacitors may be mounted 'piggy-back' over the IC chips. The inputs of a digital circuit will be closely defined, and for modern circuits are usually the levels 0V or +5V, called *TTL levels*. The outputs of a digital circuit will also be at either of these levels.

Two terms that are very important to the operation of digital circuits are *source* and *sink*, applied to the current levels at an input or output of a digital IC. When we say that an output can source 15mA, for example, we mean that the output can provide this maximum amount of current to whatever is connected to it. The connections will either be the inputs of other digital ICs, discrete circuits, or other loads such as micro-motors.

The ability to *sink* current means being able to allow inward current flow without an unacceptable change in voltage level. If we say that the input of an IC requires a sink of 1mA, for example, we mean that 1mA can flow out of the input, and this should not change the input voltage to such an extent that it will cause the level to change from 0 to 1 or from 1 to 0.

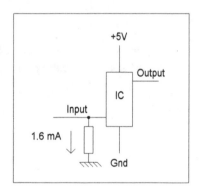

Figure 1.2   Sourcing and sinking currents

Many digital IC inputs will pass *negative* current (from the input to the circuit that drives the input) when they are at level 0, but no detectable current when they are at level 1. At level 0, the resistance between the IC input and earth must be small enough to ensure that the current will not cause a change of level. For example, the older TTL circuits required a connection that would sink 1.6mA at 0V. This restricted the resistance between the input and earth to 500Ω or less if the voltage level was not to rise to an unacceptable amount (that is, more than the maximum logic 0 level of 0.8V). Digital IC outputs can usually source (supply) or sink (absorb) equal amounts of current.

This leads to the terms *fan-out* and *fan-in*, of which fan-out is much more common. The fan-out of a digital IC means the number of standard gate inputs that can be reliably driven from the output. For example, if a gate output can supply 16mA, and its input requires sinking of 1.6mA, then one output can be connected to as many as ten inputs without compromising the ability of the circuit to operate. Fan-in is much less important, and seldom quoted; it is defined as the number of inputs to a digital circuit.

Threshold level is another important term that relates to the changeover from level 0 to 1 or from level 1 to 0. If we consider an IC that operates with TTL signals whose nominal voltage levels are 0V and +5V, the signals that are acceptable at an input need not be *exactly* at these levels. The 0 input level can extend from 0V to +0.8V and the 1 input level can be anything from +2V to +5.5V. The values of +0.8V and +2V are the threshold levels, and any input between these levels is indeterminate — you cannot predict what the output will be. Digital circuitry must therefore be designed so that these intermediate levels exist only for a very short time during the change from 0 to 1 or 1 to 0.

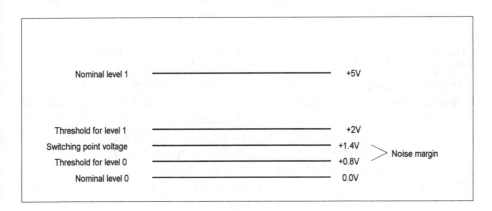

Figure 1.3   Threshold levels and noise margin

The noise margin of a digital circuit is the amplitude of unwanted input pulse that will just make the device switch over. This refers mainly to an input at the 0 level. For example, if the maximum possible input voltage level for logic level 0 is 0.8V, and the gate will actually switch when the voltage level reaches 1.4V (a typical figure), then a positive-going noise pulse of 0.6V peak amplitude will be enough to cause the gate to switch over. This value of 0.6V is the noise margin (see Figure 1.3). It is more accurately described as the noise margin for logic 0, but since the noise margin for logic 1 is usually larger, this is the value that is normally quoted.

Note that noise margins are improved if digital circuits can be operated from higher voltages, so that the typical input levels of 0V and +15V that are used by the 4000-series of CMOS chips permit a much larger noise margin. The usual TTL type of circuits, however, are restricted to a +5V supply, and noise problems are reduced by using a well-stabilised power supply with decoupling at each chip, and by using low-impedance inputs. Modern microprocessor (and allied) chips often use MOS circuitry with a 3.3V supply line to reduce dissipation, particularly for portable computers.

Scale of integration

The scale of integration is concerned with how many devices can be packed on to one IC chip. For the sake of comparison, this is given in terms of the equivalent number of simple gates, and rather than quote the numbers precisely, letters are used to indicate the range of packing. The earliest IC devices were described as small-scale integration (SSI), meaning the range 3 to 30 gates (or equivalent) per chip. This was rapidly developed to medium-scale integration (MSI), meaning 30 to 300 gates per chip, and to large-scale integration (LSI), 300 to 3,000 gates per chip.

The pace of development proceeded so rapidly that the term very large-scale integration (VLSI) had to be introduced to mean devices containing 3,000 or more gates per chip, and this term is still used when the chip contains 200,000 or more gate equivalents. In general, TTL and CMOS gate and flip-flop circuits are SSI or MSI, and most of the chips used in computing are VLSI types. Another way of referring to LSI and larger integration is in terms of the bits of memory on a memory chip of the same scale. This puts LSI at up to 16K, VLSI from 16K to 1M, and a new category, ELSI (extra large scale of integration) at 1M and above.

The most familiar chip package for logic circuits is the dual-in-line (DIL) type using 14 or more pins, illustrated for the type 7400 in Figure 1.4. DIL can be used in two versions of different widths, and packages using up to 40 pins are commonplace; a few 64-pin DIL packages have been used. The smaller packages are often illustrated with either the gate terminals, as shown, or gate drawings inside the outline to show the connections.

VLSI and ELSI packages for modern microprocessors and allied chips need a large number of pins, and square packages are now common. These may use surface-mounting or pin-insertion, and the usual way of locating pin 1 is to find the chamfered corner, Figure 1.5. The pin 1 position, looking from the top, is the first pin clockwise. When several rows of pins are used, a row-column location system is used with both numbering and lettering. The chamfered edge in this case identifies row A and column 1.

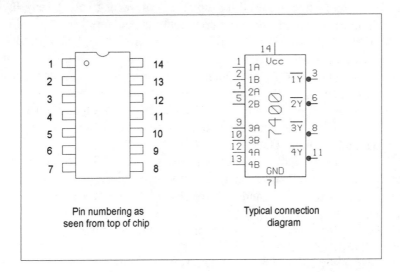

Pin numbering as
seen from top of chip

Typical connection
diagram

Figure 1.4   Typical DIL pin arrangement and connection diagram. The connection diagram shows the 4 input (A and B) and outputs (Y) for this NAND chip

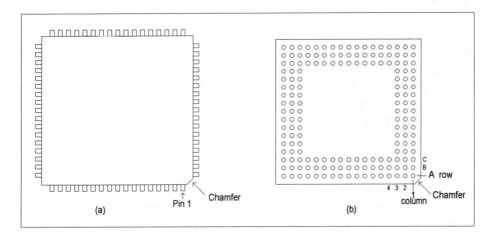

Figure 1.5   Typical 68-pin (a) and 168-pin (b) packages, showing pin identification systems

Chip circuits

The types of circuits that are used for digital ICs allow us to assign any IC to one of a few families, of which the most commonly encountered are STTL, LSTTL, ALSTTL, CMOS, HCMOS, ECL, IIL (also written as I^2L), PMOS and NMOS. These abbreviations are defined in the following pages. Though all of these families contain chips that perform identical *actions*, such as a four-input NAND gate, the way that the internal circuitry is designed can make a considerable difference to power consumption, fan-out, speed of operation and other important factors. To see why there should be such differences we need to look at some typical input and output circuits.

NOTE: Some of the families listed above, such as STTL, are almost obsolete but they remain in production because of the need to service older equipment. Other families, such as ECL and IIL, are restricted to a few specialised devices and not available for a large range of logic chips. It is usually possible to replace a chip from an older family by a more modern type, but only if you are aware of the internal differences, hence the importance of knowing something of the circuitry. It is not necessary to know details of the construction of every possible gate, only to know in general terms how the input and output circuits are constructed. Remember also that a circuit for any part of an IC is approximate only, it does not take into account stray capacitances and leakage resistance values within the chip.

STTL

TTL is an abbreviation meaning *transistor-transistor logic*, used to distinguish this type of circuit from earlier varieties of logic gate circuits such as DTL (diode-transistor logic) and RTL (resistor-transistor logic). The use of the term STTL for these circuits, with S meaning 'Standard', is more recent. The distinguishing feature of STTL circuits is the use of several bipolar transistors in integrated form, along with an input circuit which is connected to the emitter of a transistor.

Figure 1.6 shows a circuit diagram that approximates to the internal circuitry of a TTL gate of the 74 family, whose type numbers start at 7400 and continue into five and six-figure numbers. The input stage consists of a transistor which has been formed with two emitters. This is comparatively simple to carry out when an IC is being formed, and 15 or more emitters can be formed on a single base of a transistor. The output stage makes use of the familiar 'totem-pole' type of circuit. The stages between these two will implement the gate action, and they need not concern us.

Since this is a TTL circuit, it is designed to operate with a +5V stabilised supply. Any voltage level above about 2.4V at the input will be taken as being at logic level 1, and any voltage less than about 0.8V at the input will be taken as being at logic level 0.

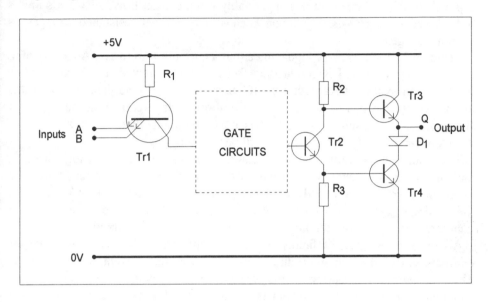

Figure 1.6   Typical input and output circuitry for STTL devices

Bearing these quantities in mind, we can now look at the operation of the circuit in detail. Imagine that the collector of Tr_1 is connected directly to the base of Tr_2. When both inputs, A and B are connected to logic 1, ideally +5V, then no current will flow between the base of Tr_1 and either emitter, since the base voltage is not at a level sufficient to pass current (0.6V above the emitter voltage). With no current passing between the base and either emitter, the collector voltage of Tr_1 will be high, with current flowing from the base of Tr_1 to the collector of Tr_1 and so into the conducting base of Tr_2. In this condition, both Tr_2 and Tr_4 will be conducting, and the output voltage at Q will be low because Tr_4 is conducting and Tr_3 is cut-off.

Imagine now that either one of the emitters is connected to logic 0. Current will flow through the base of Tr_1 to the emitter which has been connected to logic 0, and because this current is large enough to saturate the transistor, the collector voltage will be low, and no current will flow into the base of Tr_2, which will now be cut off. In this condition current flows through R_2 into the base of Tr_3, causing this transistor to conduct and thus connecting the output Q to the +5V line. Tr_4 is cut off, and the output Q is at about 3.8V, logic level 1.

Since the output at Q is low only when both inputs are at logic 1, the action is that of a NAND-gate, and the use of Tr_4 to connect the output to logic 0 ensures that comparatively large currents can pass from the terminal Q to the logic 0 line

7

without raising the voltage level of the output above the guaranteed level 0 figure. Remember that when a transistor is saturated, its collector-emitter voltage is low, typically 0.2V, and does not rise appreciably when current flows — Ohm's law is not obeyed by a transistor junction because the internal resistance changes as current changes.

Standard TTL chips carry a guarantee that a current of up to 16mA can be sunk at the output when Tr4 is conducting. Because the current that Tr3 can pass is limited by the value of R2 within the IC, there is generally no guarantee that this amount of current can be sourced because of the wide tolerance on the values of IC resistors, and designers are not advised to use circuits that require a source of 16mA of current at Q, current which will pass through Tr3.

The input currents flow *only* to the emitters that are connected to logic level 0, and the value of resistor R1 will pass a current of 1.6mA maximum. Since the maximum current (guaranteed) which can be sunk at an output is 16mA, ten times the maximum current that must be sunk at the input to hold the input level at a guaranteed logic 0 level, the fan-out of a circuit of this type is fixed at 10.

All members of the 74 family of STTL chips use inputs that are transistor emitters, so that they pass current at logic level 0 and no current at logic level 1. The propagation delay, which is the time between changing the level at the input and finding a change at the output is in the range of 11ns to 22ns ($1ns = 10^{-9}$ seconds). The power dissipated per gate under average switching conditions is 10mW, and the typical operating frequency is around 35MHz.

Low-power Schottky (LS)

At the time of writing, standard STTL is manufactured only for replacement purposes, though millions of TTL chips are still in use. Another type of TTL circuit, the low-power Schottky (LS) has in the past replaced the standard variety because of the twin advantages of high speed of operation and lower power dissipation. Using the normal STTL circuitry, high switching speed (low propagation delay) can be achieved only by passing large currents through the transistors.

Low propagation delays that are obtained by using large currents will inevitably lead to larger chip dissipation, because the current is flowing through the integrated components which are all part of the IC. The LS range of TTL ICs avoids this difficulty by using a different principle, relying on the use of Schottky diodes, components which can be made easily in integrated circuit form.

The Schottky diode, whose symbol is illustrated in Figure 1.7, is formed by evaporating aluminium on to silicon, and its remarkable feature is its very low forward voltage when it is conducting, of the order of 0.3V.

This feature is used in two ways. One application is to carry out the logic action using the diodes in circuits similar to those used in the early DTL (diode transistor logic) ICs.

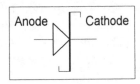

Figure 1.7   The symbol for a Schottky diode

The other use is in preventing the transistors in the circuit from saturating. A transistor is saturated when it is fully conducting, with the base passing more current than is needed to make the collector circuit conduct fully, since the collector current is limited by the value of collector load resistor. In this state, the collector-to-emitter voltage will be low, about 0.2V, as compared to the 0.6V which will exist between the base and the collector.

When a transistor is saturated, there will be a comparatively large amount of slow-moving charge in the base layer, and when the transistor is switched off by connecting the base terminal to the voltage level of the emitter, this stored charge will permit current to flow between the collector and the emitter for a time which can be as long as a microsecond. This restricts the speed at which a switching circuit can be operated, but if the transistors in switching circuits are not allowed to saturate, a very considerable increase in switching speed is possible. This cannot, however, be achieved by normal biasing methods, particularly within an IC. The other advantage of using Schottky junctions is that they operate using majority carriers, so that the storage effect does not apply.

Figure 1.8 shows the input circuit of a NAND-gate of the 74LS family which

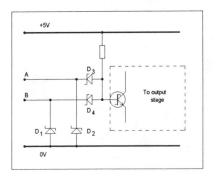

Figure 1.8   A typical LS input stage, using Schottky diodes and transistors

uses Schottky diodes for logic, and also uses diodes of the same construction connected between the collector and base of each transistor to prevent saturation

The Schottky connection within the transistor is between the base and the collector so that when the transistor collector voltage approaches the saturation value, current from the base circuit is diverted by the Schottky diode into the collector circuit. This prevents the base current from reaching the value which would cause saturation. The Schottky symbol is shown on the transistor base to make it clear that the Schottky diode is used within the transistor. The transistor circuit which is used within these Schottky TTL ICs is designed to make use of current stabilisers in addition to other methods of preventing saturation

The 74LS series operates at the same voltage levels as the STTL 74L series, but the maximum output current is 8mA, and the input current at logic level 0 is –0.4mA, one quarter of the STTL current, so that the fan-out is 20. The propagation delay is in the range 9ns to 15ns, appreciably lower than that of STTL, and the average power per gate is about 2mW. The typical switching frequency is 40MHz.

Open-collector gates

Though most of the gates of the 74 and 74LS series of TTL digital ICs make use of the 'totem-pole' type of output circuit which has been illustrated, a few gates use what are called 'open-collector' outputs. This means that the phase-splitter and one half of the output stage is missing, so that the output terminal of such a gate is simply the collector terminal of a transistor (Figure 1.9). If this gate is to be used in normal logic circuit applications, a load resistor, connected between the output and the +5V supply line, must be added externally.

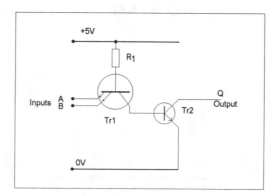

Figure 1.9   **An open-collector gate which requires a load resistor**

When the transistor, of which this resistor is a load, switches off, the voltage at the collector will rise. This rise of voltage is comparatively slow, however, because the stray capacitance at the collector is being charged through the load resistor rather than through the very much lower resistance of a transistor as it would be if the totem-pole circuit were used. The rise time of a pulse when stray capacitance is being charged through a load resistor is therefore noticeably slower than the fall time when the stray capacitance is discharged by the conducting transistor.

The advantage of using open-collector outputs is that they can make a 'wired-OR' connection, which is impossible when the totem-pole type of output stage is used. A wired-OR connection is obtained when the outputs of two open-collector gates are connected together, and to a load resistor (Figure 1.10).

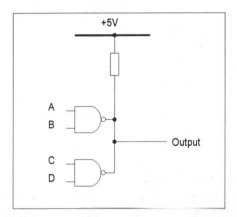

Figure 1.10   Open-collector gates connected to make a wired-OR output

In this example, the output is logic 0 when either (or both) gate output are at logic 0, so that the gate action is (A NAND B) OR (C NAND D). If this connection were made using normal gate ICs with totem-pole outputs then it would lead to the destruction of the gates when any one gate had inputs which caused a 0 output and the other had inputs that caused a 1 output. Gates of the usual totem-pole type must *never* have their outputs directly connected.

Emitter-coupled logic (ECL)

Emitter-coupled logic ICs make use of switching currents between two bipolar transistors, rather than switching transistors completely on or off. The circuits are designed so that the transistors never become saturated, thus ensuring that very high-speed switching can be obtained. This system is also known as *CML*, meaning current-mode logic, and is used when very low propagation delays are required.

11

Figure 1.11 shows the circuit of a typical ECL OR-gate, in which the input transistors Tr1 and Tr2 form one half of a balanced circuit, sharing an emitter-resistor R_1 with Tr3.

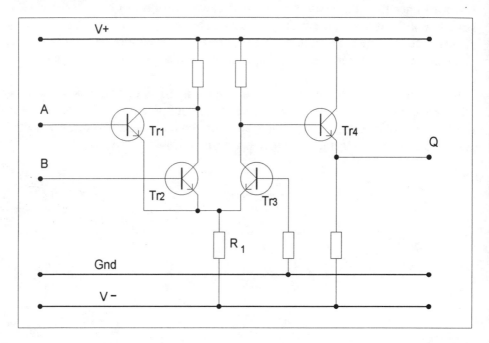

Figure 1.11 **A typical ECL circuit, showing the use of balanced pairs of transistors. The voltage levels are generally lower than for TTL, and ECL chips are usually expensive compared to their TTL equivalents**

The inputs are connected to the bases of the transistors, and if either input is high or if both inputs are high, the current through R_1 will be provided by Tr1, Tr2 or both, rather than by Tr3. This will leave only a small current flowing through Tr3, so that its collector voltage will be high. This, in turn, ensures that Tr4 conducts, making the output at its emitter terminal high.

ECL methods are not commonly used for simple gate circuits, but National Semiconductor manufacture the F100K series (with type numbers ranging from 100101 to 100182) which have delay times of less than 1ns. ECL is more commonly used for high-speed counters and on Programmable Array Logic (PAL) chips which can be programmed to carry out a set of logic actions. The ECL circuits can be used with very low voltage levels, typically with logic 0 level of −1.55V and logic 1 at −0.75V. If such voltage levels are used, a buffer stage will be needed to convert to normal TTL levels. Modern ECL designs can use a negative 4.5V line, with typical propagation delays of 1ns or less (not more than 6ns). The power dissipation depends on the supply voltage used, but is generally higher than that of STTL. Because the ECL stages work with small voltage differences between the logic levels, their noise margins are small. ECL counters can operate at frequencies of 500MHz and more.

IIL (or I^2L)

Integrated injection logic devices are the other example of bipolar design that feature low propagation delays, but, like ECL, they are unlikely to be used for manufacturing families of simple gates but are more likely to be part of a larger-scale logic circuit.

The basic system uses a PNP transistor to inject current into the base of an NPN transistor, and switching this current either into an output stage (to make the output low) or to an input stage (allowing the output to remain high).

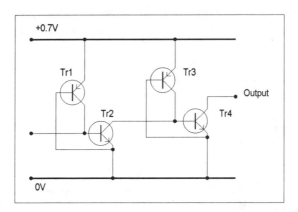

Figure 1.12   Principles of I^2L circuits. Only one collector has been shown on Tr2 and Tr4

In the example of Figure 1.12, with the input at logic 1 of 0.4V, Tr1 is shut off and so also is Tr2. The current passing through Tr3 maintains Tr4 switched on, with its collector passing current. When the input is taken low, Tr1 and Tr2 will be switched on, so that the current from Tr3 flows into Tr2 and Tr4 cuts off, changing its logic state. As for ECL, buffer stages are needed to interface with TTL voltage levels, so that I^2L is seldom used in conjunction with with TTL devices.

Devices are normally manufactured with more than one collector junction so that one I^2L device can drive several other inputs — the multiple collectors provide the fan-out for the circuit. The voltage swing between high and low is

13

small, typically 0.7V, so that the noise margin is low. Packing density is high, and the propagation delay is lower than that of TTL, so that I²L devices are relatively fast in operation. Power dissipation is around the same as that of STTL, and it is possible to choose between fast operation and low power by adjusting the injection current. At the time of writing, very few I²L devices are being manufactured, and there is no I²L family that corresponds to the TTL 7000 series.

CMOS gate circuits

NOTE: *Before* you read this section, you should revise the principles of MOSFETs, see Volume 2, Part 1.

By far the most common semiconductor technology in current use is that of field-effect transistors (FETs). Three types of ICs can be manufactured using P-channel and N-channel FETs. PMOS ICs use P-channel FETs exclusively, NMOS ICs use N-channel FETs exclusivel, and CMOS (C meaning complementary) ICs make use of both P- and N-channel FETs in a single circuit. PMOS methods were initially used for manufacturing microprocessors and similar chips, but were superseded first by NMOS and later by fast versions of CMOS.

A typical CMOS circuit, that of one NAND-gate of the CD4011A quad NAND IC, is shown in Figure 1.13 In this circuit, Tr1 and Tr2 are both P-channel types; Tr3 and Tr4 are N-channel types. The P-channel FETs will be switched into conduction by a logic 0 input at their gates, since their sources are connected to the positive supply.

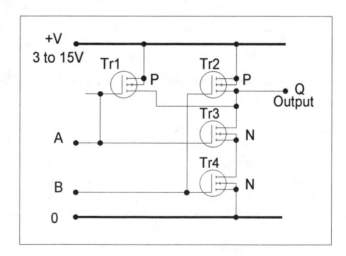

Figure 1.13   A typical CMOS circuit, in this example a NAND gate

The N-channel FETs will be switched into conduction by a logic 1 input at their gates, because their sources are connected to the 0 voltage line. With either or both gate inputs at logic 0, the P-channel FETs will conduct, keeping the output high. Only when both inputs are high can both N-channel FETs conduct, and thus connect the output to the logic 0 level. The action is therefore that of a NAND-gate. A NOR gate can be created using the same set of components by connecting the N-channel FETs in parallel and the P-channel FETs in series.

These CMOS ICs can operate with a wide range of supply voltages, typically 3V to 18V, and with very small currents flowing, typically 5μA. The logic 0 and 1 voltages are normally very much closer to the supply voltage levels than is possible with bipolar designs of the TTL, ECL or I^2L types. For example, using a +5V supply, a logic 1 voltage of +4.95V and a logic 0 voltage of 0.05V can be obtained.

The input current is always negligibly small because the inputs are connected to FET gates, and the output currents are typically about 0.5mA maximum. The fan-out figure for low-frequency operations can be very large, 100 or more, but the value decreases as the frequency of operation is increased. This is because the small currents that are available at the output must be capable of charging and discharging the capacitance at the input of each gate which is connected to the output. This requirement for charging and discharging stray capacitances also increases the total dissipation of the IC as the frequency of operation is increased.

A simple gate, for example, which has a dissipation of 1μW at a frequency of switching of 1kHz, may have a dissipation of 0.1mW at an operating frequency of 1MHz. This factor limits the operating speed of CMOS circuits, and leads to the earlier types CMOS ICs (the 4000 series) being used in low-speed applications rather than for high-speed machine-control or computing applications. They are widely used where speed is not of primary importance, though.

The very high insulation resistance of the gates makes them very susceptible to damage from electrostatic charges, and modern CMOS ICs are manufactured with a network of diodes connected to the inputs which will conduct whenever the voltage between gate and source or gate and drain becomes excessive. These diodes will protect for static voltages of up to 4kV, but if higher voltages are likely to be encountered stringent earthing precautions must be taken. For example, operators may be required to use metal wrist straps that are earthed, and work on a conducting earthed surface. The safest way to work with CMOS devices is to earth all pins together until they are inserted into place and connected. Note that walking along a nylon carpet can generate voltage levels in excess of 16kV.

CMOS families originally used numbering from 4000 upwards, and the numbers were later supplemented with the letters BE (older types) and UBE. These letterings mean, respectively, buffered and unbuffered. Later, new families

15

of CMOS used the numbering of TTL devices, with the letters HC, HCT or AC, so that a 74HC00 was a gate equivalent to the 7400 or 74LS00 in action, but with greatly reduced power consumption. These CMOS equivalents can be used, subject to some caution on their characteristics, as replacements for bipolar TTL devices.

The older 4000BE series have typical propagation delays of 125ns to 250ns, power dissipation per gate of 0.6μW, and a typical switching frequency of 5MHz. The UBE family feature propagation delays of 90ns to 180ns and slightly higher typical frequency ratings. The 74HC family are direct replacements for 74LS types, with propagation times of 8ns to 15ns, power dissipation of 1μW, and typical frequency of 40MHz. The 74HCT is very similar, but with slightly longer propagation delays. The most recent CMOS family is designated 74AC, with propagation delays of 5ns, power dissipation of 1μW, and a typical frequency of 100MHz.

NMOS and PMOS

Devices can be constructed using only one type of FET structure, and these are designated as NMOS and PMOS types. Neither is used to create families like the 7400 or CMOS 4000 series, but both are extensively used for microprocessor chips and for associated chips. PMOS devices are comparatively easy to fabricate in VLSI sizes, but are slow compared to NMOS, so that PMOS is less common now, but NMOS was used from the 1980s onwards as a manufacturing system for microprocessors. The most recent chips use HCMOS because of its superior speed and low power consumption.

Table 1.1 shows a comparison of the most common families of devices that use either the 7400 or the 4000 type numbers.

Table 1.1   IC families summarised

	STTL	LSTTL	CMOS	74HC	74AC
V+supply	5V	5V	3–15V	5V	5V
Imax/1	40μA	20μA	10pA	10pA	10pA
Imax/0	−1.6mA	−0.4mA	10pA	10pA	10pA
Imax/out	16mA	8mA	1mA*	4mA	4mA
Delay	11–22ns	9–15ns	40–250ns*	10ns*	5ns
Power	10mW	2mW	0.6μW	1μW	1μW
Frequency	35MHz	40MHz	5MHz	40MHz	100MHz

NOTES:

V+supply = normal positive supply voltage level

Imax/1 = maximum input current for logic level 1

Imax/0 = maximum input current for logic level 0   (*continued opposite*)

Imax/out = maximum output current
Delay = propagation delay in nanoseconds
Power = No-signal Power dissipation per gate in mW or µW
Frequency = Typical operating frequency
$1pA = 10^{-12}A$
* These quantities depend on the supply voltage level.

Exercise 1.1

Connect a logic gate as illustrated in the diagram, Figure 1.14, using an STTL 7400 chip initially. Monitoring the output level, gradually raise the DC input level from zero to find the point where the output level just changes over. Note this voltage. Starting from logic level 1, gradually reduce the steady input voltage until the chip switches over again, and note this voltage level also. Draw a diagram to show the threshold levels. Repeat with 74LS00, the CMOS 4011 and 74HC00 and any gates of the IIL and ECL types that can be obtained.

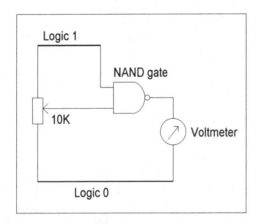

Figure 1.14   A circuit for investigating threshold levels for a gate

Exercise 1.2

Use a NAND gate in the circuit of Figure 1.15 and measure the rate at which the output changes. What is the output waveform? Use this circuit with as many different types of NAND gate as you can find, such as STTL, LSTTL, CMOS 4000, HC etc. For one type of family, such as LSTTL, find the relationship between the RC time constant and the frequency of oscillation.

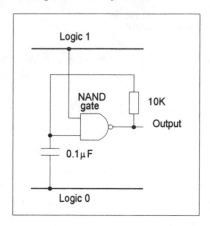

Figure 1.15   An oscillator circuit that uses an inverting gate

Test Questions

1.   If a chip circuit can provide a current of 1.6mA at its output and requires to sink a current of 200μA at an input, what is its fan-out value?

2.   Why can you not quote a single value of fan-out for a CMOS circuit?

3.   A digital counter circuit is to be constructed to operate at high frequencies in excess of 500MHz. What type of chip technology should be used for the first stage (at least)?

4.   What do you think would be the most appropriate technology for designing a pocket calculator, and why?

5.   For some industrial controller circuits, CMOS chips working with a 15V supply are preferred to TTL using +5V supplies. Why?

6.   You need to replace some gates in a circuit, and you find that the outputs are connected. What type of gate design must be used as replacements?

7.   What values of power per gate and propagation delay would you expect in a device marked as 4010?

8.   What values of voltage would you take as being acceptable for logic 1 state in a 74 series TTL gate output with the usual +5V supply?

9.   What scale of integration would you expect to be used for (a) a Quad NAND-gate, (b) a PIO, (c) an 8-bit microprocessor?

10.   Name the logic families corresponding to type numbers 4001, 74LS240, 74HC10, 7400.

2   Number systems, arithmetic, error control and code formats

Syllabus references: D02–2.2, M01–1.1, M01–1.2, M01–4, B2–C.6, D07–7.4, M01–1, I2.2, B1–E.5, B2–C.6, D07–7.4

Digital arithmetic

The basic principles of arithmetic were covered in Volume 1 of this series, to which the reader is referred for any necessary revision. Nearly all electronic computing and logic systems operate on a binary base or radix due to the on-off nature of the signals generated from a switching action. Due to using such a small radix, any reasonably large number requires a long string of symbols to represent it. For example, with an 8-bit word, the maximum quantity that can be represented is $2^8 = 256$ different symbols, i.e. 0 to 255. By extending the binary system to include two 8-bit groups or bytes per number, it becomes possible to use an extension of the standard form representation for very small and very large numbers. In this case the second byte is referred to as the power or exponent of the number. For example, $300,000,000 = 3 \times 10^8$, where 10^8 is described as the exponent part of the number. This is usually referred to as floating point arithmetic.

Binary coded decimal (BCD)

This form of coding is commonly used when numbers need to be continually displayed in their decimal format, whilst the processing is being carried out in binary. Each denary digit is represented by its 4-bit binary equivalent as listed in Table 2.1. In practice, the interchange between each format is performed by encoder/decoder circuits. By convention, the right-most bit which has the lowest weighting is described as the least significant bit (LSB). It therefore follows that the left-most bit with the highest weighting is described as the most significant bit (MSB).

Table 2.1   BCD codes

Denary	BCD	Denary	BCD
0	0000	1	0001
2	0010	3	0011
4	0100	5	0101
6	0110	7	0111
8	1000	9	1001

Gray codes

Reference to Table 2.1 will show that the binary codes for successive denary numbers change by several bits at each step. For example, the step from 7 to 8 involves a change of all 4 bits. Under electrically noisy conditions, this can easily lead to errors. The set of Gray codes which were developed to overcome this particular problem, only involves 1 bit change at each step. If two or more bit changes are detected, then an error must have occurred. One example of a 4-bit cyclic Gray code is shown in Table 2.2, where the bit pattern changes in a revolving fashion one bit at a time, repeating every 8 or 16 steps.

Table 2.2   Denary and Gray codes

Denary	Gray	Denary	Gray	Denary	Gray	Denary	Gray
0	0000	1	0001	2	0011	3	0010
4	0110	5	0111	6	0101	7	0100
8	1100	9	1101	10	1111	11	1110
12	1010	13	1011	14	1001	15	1000

Signed numbers

Just as the denary system can be used to represent positive, negative and fractional numbers, so can the binary system. In this case it is common to use the leading or MSB in the binary stream to represent the polarity (positive or negative) of the number. By convention, a leading zero signifies a positive number. Thus the 4-bit word, 0101 is +5, whilst 1101 is –5. Computer systems commonly operate with 8-bit bytes, so that using this concept, operation is restricted to $-2^7 = -128$ to $+2^7 = +128$. Since zero is considered to be the first positive digit and appears in both halves of the sequence, this concept covers the range of decimal numbers from –127 to +127. However, this range can be considerably extended by using the floating point concept described above.

Fractional binary numbers

Binary numbers have a fractional form that is represented in a manner similar to that used for decimal numbers. The decimal point or binary point is again used to separate the integer and fractional parts of the number. In this case, the weighting of each bit to the right of the binary point is based on powers of –2 as indicated in Table 2.3.

Table 2.3   Powers of –2

Power or exponent	2^1	2^0	2^{-1}	2^{-2}	2^{-3}
Place weighting	2	1	0.5	0.25	0.125

For example, $10.101 = 2.(0.5 + 0.125) = 2.625$

Other coding methods

It was pointed out above, that to represent any reasonably large decimal number required a long string of binary bits. If a human operator has to enter a large amount of data into a computer system in this form, it will take a long time. In addition, the tedium will lead to operator errors. What is needed is a short code for the operator to use and one which the computer can easily convert into the binary format by itself.

Octal coding

If a string of binary bits is divided into groups of three bits, each group can be used to represent one of $2^3 = 8$ different symbols. The relationship between octal and decimal coding is shown in Table 2.4, where it will be seen that this format is the same as the first eight entries in the BCD Table. Using this table, the binary

Table 2.4   Binary and octal codes

Binary	Octal	Binary	Octal
000	0	001	1
010	2	011	3
100	4	101	5
110	6	111	7

string 111,010,101 can be represented by 725 and vice versa. To distinguish such an octal number from a decimal one it is common to add a suffix 8, that is 725_8. With this code format, the number of symbols for data entry is reduced and the computer can easily be programmed to convert this into binary.

Exercise 2.1.

1. To convert the denary number 298 into both binary and octal. By repeated division by 2, $298_{10} = 100\ 101\ 010$ and in octal format this becomes 452_8.
2. Convert the octal number 762 into both binary and decimal. By direct exchange from Table 2.4, 762_8 becomes 111 110 010.
   $$762_8 = 7\ x\ 8^2 + 6\ x\ 8^1 + 2\ x\ 8^0 = 448 + 48 + 2 = 498_{10}.$$

Hexadecimal coding

Because an 8-bit byte does not divide exactly into groups of three bits, octal coding is seldom used for computer driven systems, instead an hexadecimal format which is based on $2^4 = 16$ is preferred. An 8-bit byte can be represented by two hexadecimal (Hex or H) symbols. As indicated by Table 2.5, the hexadecimal symbol set uses the decimal characters 0 to 9 plus the upper-case letters A to F. This table should also be compared with those for both BCD and octal coding.

Table 2.5   Decimal, binary and hexadecimal codes

Decimal	Binary	Hexadecimal	Decimal	Binary	Hexadecimal
0	0000	0	1	0001	1
2	0010	2	3	0011	3
4	0100	4	5	0101	5
6	0110	6	7	0111	7
8	1000	8	9	1001	9
10	1010	A	11	1011	B
12	1100	C	13	1101	D
14	1110	E	15	1111	F

To convert from binary into hexadecimal, each byte is divided into groups of 4 bits, to the left and right of the binary point and each then replaced by the corresponding character from Table 2.5. For example, 1101 0100 becomes D4H, (where H signifies that the symbols D4 represent a hexadecimal number); whilst 100011.011111 becomes 0010 0011.0111 1100 by adding leading and trailing zeros to make up groups of 4 bits, which in turn converts to 23.7CH.

The weighting of each symbol and its hexadecimal equivalent are shown for up to four character hexadecimal numbers in Table 2.6.

Table 2.6   Hexadecimal to decimal converter

Hex digit		Exponent	values	
	3	2	1	0
0	0	0	0	0
1	4096	256	16	1
2	8192	512	32	2
3	12288	768	48	3
4	16384	1024	64	4
5	20480	1280	80	5
6	24576	1536	96	6
7	28672	1792	112	7
8	32768	2048	128	8
9	36864	2304	144	9
A	40960	2560	160	10
B	45056	2816	176	11
C	49152	3072	192	12
D	53248	3328	208	13
E	57344	3584	224	14
F	61440	3840	240	15

Exercise 2.2.

$2EH = 2 \times 16^1 + E \times 16^0 = 32 + 14 = 46_{10}$

$A4H = A \times 16^1 + 4 \times 16^0 = 160 + 4 = 164_{10}$

$B4A2H = B \times 16^3 + 4 \times 16^2 + A \times 16^1 + 2 \times 16^0 =$

$\qquad 11 \times 4096 + 4 \times 256 + 10 \times 16 + 2 \times 1 =$

$\qquad 45056 + 1024 + 160 + 2 = 46242_{10}$

$A8H = 1010\ 1000 = 168_{10}$

$216_{10} = 1101\ 1000 = D8H$

Multiple units

As with the decimal scale where the letters k, M and G are often used to abbreviate the multiple values of 10^3, 10^6 and 10^9, these are also used with binary quantities but in the following manner:

$$k \text{ (kilo)} = 2^{10} = 1024, = 1.024 \times 10^3$$
$$M \text{ (Mega)} = 2^{20} = 1048576 = 1.048576 \times 10^6$$
$$G \text{ (Giga)} = 2^{30} = 1073740000 = 1.07374 \times 10^9$$

Often terms such as Mb and MB are used to signify megabit and megabyte respectively. However misprints can occur that create significant errors and so Mbit and Mbyte are preferred.

Addition

The rules for binary addition can be stated as follows;

$0 + 0 = 0$ $\qquad\qquad 0 + 1 = 1$ $\qquad\qquad 1 + 0 = 1$ $\qquad\qquad 1 + 1 = 0$ Carry 1

Thus two binary numbers can be added as follows;

$$01011010$$
$$01101011$$
$$\overline{11000101}$$

The rules for hexadecimal addition are best stated in tabular form.

Table 2.7   Adding hexadecimal numbers

0	1	2	3	4	5	6	7	8	9	A	B	C	D	E	F
1	2	3	4	5	6	7	8	9	A	B	C	D	E	F	10
2	3	4	5	6	7	8	9	A	B	C	D	E	F	10	11
3	4	5	6	7	8	9	A	B	C	D	E	F	10	11	12
4	5	6	7	8	9	A	B	C	D	E	F	10	11	12	13
5	6	7	8	9	A	B	C	D	E	F	10	11	12	13	14
6	7	8	9	A	B	C	D	E	F	10	11	12	13	14	15
7	8	9	A	B	C	D	E	F	10	11	12	13	14	15	16
8	9	A	B	C	D	E	F	10	11	12	13	14	15	16	17
9	A	B	C	D	E	F	10	11	12	13	14	15	16	17	18
A	B	C	D	E	F	10	11	12	13	14	15	16	17	18	19
B	C	D	E	F	10	11	12	13	14	15	16	17	18	19	1A
C	D	E	F	10	11	12	13	14	15	16	17	18	19	1A	1B
D	E	F	10	11	12	13	14	15	16	17	18	19	1A	1B	1C
E	F	10	11	12	13	14	15	16	17	18	19	1A	1B	1C	1D
F	10	11	12	13	14	15	16	17	18	19	1A	1B	1C	1D	1E

Table 2.7 shows how the sum obtained by adding the head of any row to the head of any column is given at the intersection. Inversely, this table can also be used for subtraction.

Subtraction

The rules for binary subtraction can be stated as follows;

$0 - 0 = 0$ $0 - 1 = -1$ $1 - 0 = 1$ $1 - 1 = 0$

For example,

$$\begin{array}{r} 1101 \\ 0100\ - \\ \hline 1001\ \text{difference} \end{array}$$

The negative sign in the second entry of the table creates difficulties so that subtraction is better solved by the method of complements. If a number x is to be subtracted from another number y, x is first converted into an equivalent negative number and then added to y to obtain the answer. The technique can be used with any number base or radix and the following is an example using denary numbers.

By the normal method of subtraction $96 - 72 = 24$. The complement of 72 is obtained by subtracting each digit in turn from 9 to obtain 27 and adding 1 to give 28. Now $96 + 28 = 124$. Assuming that the adder system only works to two digits, the leading 1 will overflow so that the complements method yields the same answer of 24.

The 1s complement or end-around-carry method of subtraction

Using the above example

	1101
Complement 0100	1011
Now add	11000
	1
	1001

The MSB is carried round to the LSB and added to produce the same answer, rejecting the overflow bit.

The 2s complement method

The 2s complement of a binary number is obtained by inverting each bit in turn and then adding 1, i.e., find the 1s complement and add 1.

Thus the 2s complement of 0100 is 1011 + 1 = 1100 and the subtraction can now proceed via addition:

$$
\begin{array}{r}
1101 \\
1100 + \\
\hline
11001
\end{array}
$$

The MSB overflows so that this method also yields the same answer.

There is a short cut to finding the 2s complement of a binary number. Copy each bit in turn from the right hand side, up to and including the first 1. Then invert the other bits in turn, i.e.,

The 2s complement of   10011100 is thus
01100100

Method of hexadecimal complement subtraction

To evaluate A3 – 8B, we first find the complement of 8B;

8B (now find what must be added to each symbol to make 15_{10} or FFH)
74 (now add 1)
75 (the hexadecimal complement of 8B)

A3 +
75
118 (the leading symbol 1 overflows).
Thus A3 – 8B = 18(Hex).

Addition and subtraction logic circuits

Adder circuits

Reference to the rules of binary addition show that the logic gates required are a combination of Ex-Or and AND functions as shown in Figure 2.1.

However, since this circuit will only add two numbers A and B and without a Carry input, it is described as a half adder. The circuit shown in Figure 2.2 indicates how the output of two such half adders can be combined to provide a full adder to handle any Carry bit that is generated.

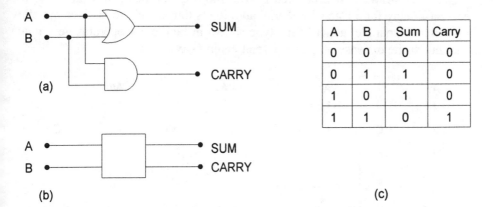

A	B	Sum	Carry
0	0	0	0
0	1	1	0
1	0	1	0
1	1	0	1

Figure 2.1   Half adder: (a) logic circuit, (b) block symbol, (c) truth table

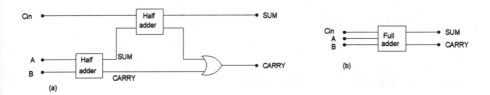

Figure 2.2   (a) Interpretation of full adder, (b) block symbol

Two binary bit streams A and B can be added in serial form if synchronised by a clock signal. Any Carry that is generated as a result of an addition needs to be added in with the next higher weighted bit pair.

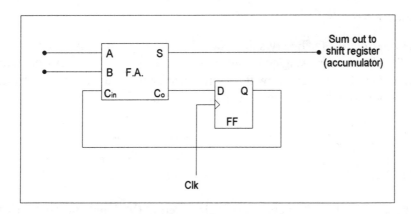

Figure 2.3   Serial adder circuit

27

Figure 2.3 shows how a clocked D-type flip-flop can be used to generate the necessary delay for a Carry bit. Serial addition is thus simple to achieve but slow in practice. A parallel adder of the type shown in Figure 2.4 operates very much faster but at the expense of greater circuit complexity.

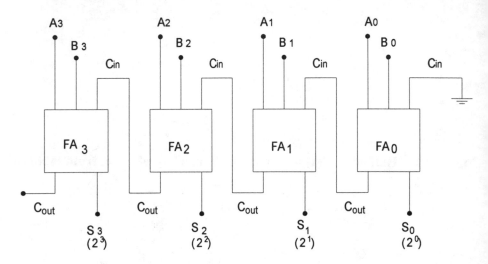

Figure 2.4   Parallel full adder circuit

All the bits of both numbers A and B are input simultaneously and any Carry signal ripples through from adder to adder as it is generated. Since addition produces no Carry at the LSB input, this stage could be replaced with a half adder. However in the interests of economy of component types, a full adder is normally used and the Carry input permanently grounded.

Subtractor circuits

In the rules matrix for binary subtraction the -1 entry represents a borrow action from the next higher weighted digit, rather than the carry forward operation of addition. Figure 2.5 shows how this feature can be implemented simply by adding an invertor to one input of the AND gate of the half adder circuit.

Figure 2.5   Half subtractor: (a) logic circuit, (b) block symbol, (c) truth table

However since this does not provide for a borrow input, the circuit is known as a half subtractor. A full subtractor circuit can be implemented from two half adders and an OR gate in the manner shown in Figure 2.6, but this again is restricted to serial operation only.

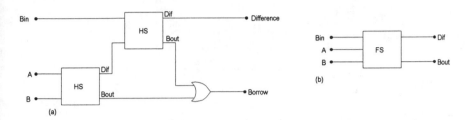

Figure 2.6 (a) Implementation of full subtractor, (b) block symbol

A much faster parallel subtractor circuit shown in Figure 2.7, can be constructed from full subtractors in the manner of the parallel adder configuration.

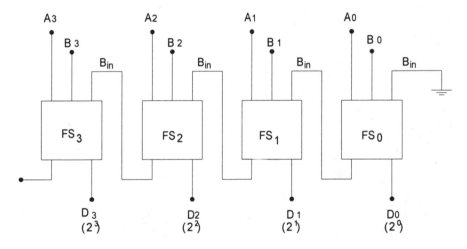

Figure 2.7 Parallel subtractor circuit

A variation of this circuit can constructed from full adder gates plus a number of invertors as shown in Figure 2.8. The feedback loop and the invertors cause this form of subtraction to be performed by the 1s complement and end-around-carry method.

29

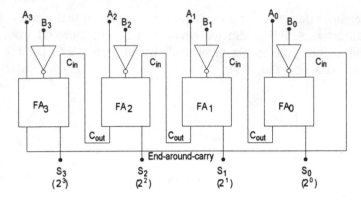

Figure 2.8   4-bit subtractor using full adders and inverters

Multiplication

The rules of binary multiplication can be stated in matrix form:

$$0 \times 0 = 0$$
$$0 \times 1 = 0$$
$$1 \times 0 = 0$$
$$1 \times 1 = 1$$

There are two ways in which this is manually carried out in practice:

```
      1101                          1101
      1010 x                        1010 x
      0000                          1101
      1101                          0000
     0000                          1101
     1101                          0000
  10000010                      10000010
```

Note that the partial product is zero if the multiplier is 0 and unchanged if the multiplier is 1. Since multiplication involves either a right or left shift and the partial products are finally summed, this technique is described as the shift and add method. The final product may be twice as long as the longest multiplying term.

Multiplier-accumulator (MAC) integrated circuits are available where a series of multiplication steps may be accumulated to form a sum of products. These are specifically designed to handle long binary bit streams, typically operating with 16 x 16 bit multipliers.

Division

Because division by zero is meaningless, the rules of binary division are simply:

$$0 \div 1 = 0 \text{ and } 1 \div 1 = 1$$

The following is an example of manual binary long division. It represents a repeated approximation process which is simulated in digital hardware by a shift and subtract process.

```
              101001
      101)11001101
          101
           101
           101
            0101
            101
            000
```

Multiplication logic circuits

Figure 2.9 indicates how a 4-bit parallel adder and an 8-bit accumulator register may be used to carry out binary multiplication of two 4-bit numbers. The multiplicand (the number to be multiplied) is loaded into one register, the multiplier into another and the accumulator cleared or set to zero. The right-most or LSB of the multiplier controls an add/not-add function in the parallel adder. The addition of the partial sums in the accumulator only occur when this bit is set to 1.

In the example shown earlier and in Figure 2.9, at the start the LSB of the multiplier is 0 so no addition occurs. At the next step, the contents of the accumulator and multiplier registers are shifted 1 bit towards the LSB. A zero is now loaded into the MSB of the accumulator so that its original contents progress through the register towards the LSB position. The LSB of the multiplier is now set to 1 so that the contents of the multiplicand register are now added to the accumulator contents. This process of shift and add continues until all the bits of the multiplier have been used. The product of the two numbers then resides in the accumulator.

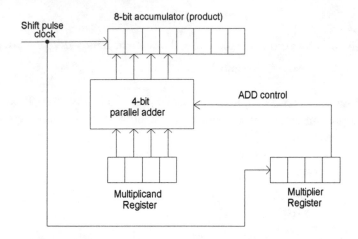

Figure 2.9   Binary multiplier shift and add circuit

Figure 2.10 shows an alternative approach to binary multiplication by repeated addition. The accumulator is first set to zero and the multiplier register value is used as a decremental counter to control the number of times the multiplicand is added to the accumulator contents. When this count reaches zero, the accumulator holds the product of the two numbers.

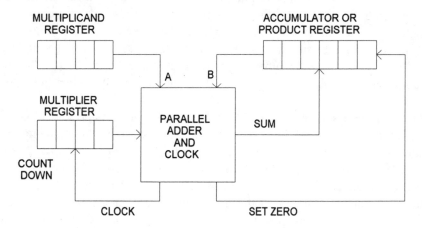

Figure 2.10   Multiplication by repeated addition

Division logic circuits

Binary division shift and subtract circuits are basically multipliers but with the inclusion of invertors. This method is very similar to the way in which adders are converted into subtractors.

Multiplication and division using a barrel shifter

The barrel shifter is a multi-register device that can be visualised as a series of registers arranged longitudinally around the surface of a cylinder. Data words can be rotated around the barrel by transfer from register to register in a parallel format. In addition, it is also possible to introduce a right or left shift to data bits to perform operations of multiplication or division. The output from the barrel shifter is via an accumulator register in which the different register contents can be added. Using a shift and rotate technique, it becomes possible to perform multiplication and division operations as shown by the following examples.

$$1010 \times 011(3) \qquad \begin{array}{l} 1010 - x\,1 \\ \underline{10100 - x\,2,} \\ 11110 - x\,3. \end{array} \qquad \text{now add,}$$

$$011 \times 101(5) \qquad \begin{array}{l} 011 - x\,1 \\ \underline{01100 - x\,4,} \\ 01111 - x\,5 \end{array} \qquad \text{now add}$$

$$0110 \times 110(6) \qquad \begin{array}{l} 01100 - x\,2, \\ \underline{011000 - x\,4,} \\ 100100 \; - x\,6 \end{array} \qquad \text{now add}$$

$$101 \times 0.11(0.75) \qquad \begin{array}{l} 101. - x\,1, \\ 10.1 - x\,0.5, \\ \underline{1.01 - x\,0.25} \\ 11.11 - x\,0.75. \end{array} \qquad \text{now add}$$

Exercise 2.3

Using a logic trainer/tutor, investigate some of the arithmetic circuits described above.

Modulo-N arithmetic

This typically involves the summation of a range of numbers, followed by division by N. The answer to the problem is then simply the remainder. For example, the sum Modulo-2 of two binary bit streams is simply the logical Ex.Or combination of the two.

Error control

Digital communications systems are more robust than the equivalent analogue systems in the presence of noise. This is clearly seen by reference to Figure 2.11(a), which shows how a clean signal can be regenerated simply by slicing the received signal at appropriate levels. Figure 2.11(b) further compares the two types of signals under the same degree of decreasing S/N ratio at the input. Whilst the analogue system degrades gracefully before it fails, the digital system is relatively unaffected by the noise until a point is reached, when the system suddenly crashes as the bit error rate (BER) rises. However, the advantages of the digital technology does not end here. There are a number of ways in which the information can be accurately recovered even from signals with a high BER.

At the simplest level, a system known as Automatic ReQuest for repeat (ARQ) is available. Reference to the ASCII (American Code for Information Interchange) code table shows that two special codes are available: ACK and NAK. If a distant receiver detects a code pattern without errors, it transmits via a return channel the code ACK (Acknowledge). However, if errors have been detected, transmission of the NAK (Negative acknowledge) code automatically generates a request for a repeat transmission of the last block of signal code.

Furthermore, bit errors can be detected and even corrected in a digital system using a technique known as forward error control (FEC). This is so called because the means of error detection/correction is contained within the transmitted message stream. This is achieved by the addition of extra redundant bits which when suitably processed, are capable of identifying the errors. Either method makes extra demands upon the spectrum; FEC requires additional time or bandwidth to include the extra bits, whilst the ARQ system requires a free return channel.

The prime causes of bit errors are White and impulsive noise. The former produces errors that are completely uncorrelated and random in occurrence,. whilst the latter creates a loss of bit stream synchronism which leads to bursts of errors.

There are three classes of error that need to be considered;
1. Detectable and correctable,

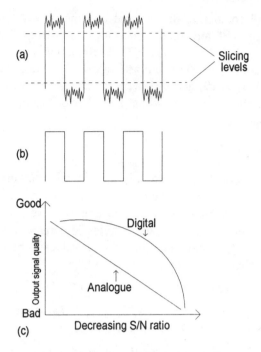

Figure 2.11 **(a), (b) Regenerating a digital signal, (c) comparison of signal behaviour in noisy conditions**

2. Detectable but not correctable, and

3. Undetectable and hence uncorrectable.

For any errors detected under Class 2, the concept of error concealment can then be applied as follows:

a. Ignore the error and treat it as a zero level,

b. Repeat the last known correct value or,

c. Interpolate between two known correct values.

Forward error control can thus significantly enhance the robustness of a digital communications system in a noisy environment. The ASCII code is a commonly used method for representing alphanumeric characters in a digital system. This 7-bit code allows for $2^7 = 128$ different alphabetic, numeric and control characters. The most commonly used digital word length is 8 bits or 1 byte, therefore there is space for one extra redundant bit in each code pattern.

Check-sum technique

This error control scheme is often used with magnetic storage media where data is stored in long addressable blocks. It is a simple scheme in which the digital sum of the numbers in any block is stored at the end of each block. Recalculating the check sum and comparing the original and the new values after each data transfer, quickly tests if any read errors have occurred. For example, if the following binary words 0101 (5), 0011 (3), 1010 (10) and 0010 (2) are valid then the check sum would be 10100 (20). If after a read operation the numbers became 0101 (5), 0011 (3), 1001 (9) and 0010 (2), the check sum should be 10011 (19). Comparison of the check sums show that a read error has occurred and the block should be read again.

Weighted check-sum technique

The simple check-sum technique can only identify when an error occurs and can not indicate the position of the actual error. However, this problem can be overcome by using a weighting scheme that employ a series of prime numbers. This can be most easily explained using a series of denary numbers. Suppose the prime numbers are 1, 3, 5, and 7 and the decimal values to be stored and read are 5, 3, 9, and 2. The check sum would calculated from:

$$1x5 + 3x3 + 5x9 + 7x2 = 5 + 9 + 45 + 14 = 73.$$

The sequence would thus be stored as 5, 3, 9, 2, 73. If on reading this became 5, 3, 8, 2, 73, the recalculated check-sum would be:

$$1x5 + 3x3 + 5x8 + 7x2 = 5 + 9 + 40 + 14 = 68.$$

The check-sums are different so an error has occurred, but the difference is $73 - 68 = 5$. Therefore a unit error has occurred in the 5 weighted value.

Even and odd parity

A single-error detection (SED) code of n binary digits is produced by placing n–1 information or message bits in the first n–1 bit positions of each word. The nth position is then filled with a 0 or 1 (the parity bit), so that the entire code word contains an even number of 1s. If such a code word is received over a noisy link and is found to contain an odd number of 1s, then an error must have occurred. Alternatively, a system might use odd parity, where the nth bit is such that the code word will contain an odd number of 1s. In either case, a parity check at the receiver will detect when an odd number of errors has occurred. The effects of all even numbers of errors is self-cancelling, so that these will pass undetected. The even or odd parity bits can be generated or tested, using Exclusive OR or Exclusive NOR logic respectively.

Such a pattern of bits is described as an (n,k) code, n bits long and containing k bits of information. It thus follows that there are n–k = c parity or protection bits in each code word. The set of 2k possible code words is described as a Block or Linear code.

The simple parity scheme which provides only a low level of error control, is suitable only for systems that operate in relatively low levels of noise.

Hamming codes

Error correcting codes have been devised and named after R. W. Hamming, the originator of much of the early work on error control. Figure 2.12(a) shows how a message may be expanded with redundant check bits.

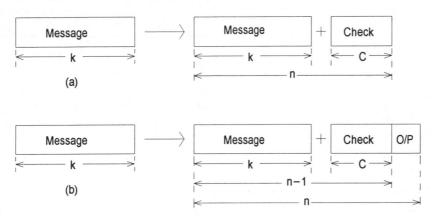

Figure 2.12 **Hamming codes: (a) single error correcting (SEC), (b) single error correcting/double error detection (DED)**

These are usually interleaved with the message bits and placed in positions 2^0, 2^1, 2^2, etc. in the encoded pattern. The mechanics for the encoding/decoding process can be explained using Table 2.8 for a (7,4) block code (block length n = 7, message length k = 4 and parity bits c = 3).

The message to be transmitted is (0011) and these bits are loaded in the 3rd, 5th, 6th and 7th positions respectively. The three parity checks are carried out, to determine the values to be placed in positions 1, 2, and 4. In general, the xth parity check bit is given by the sum modulo-2 (the remainder after division by 2) of all the information bits in the positions where there is a 1 in the xth binary position number. The transmitted code word thus becomes 1000011. If this is now received over a noisy link as 1000001, there is an error in the 6th (110) position. The receiver decoder then performs the same three parity checks and generates the following results:

1st check	0
2nd check	1
3rd check	1

The reverse of this series which is called the *Syndrome* (syndrome being a medical term for the symptoms of a disease) points to an error in bit 6 position. Now that the error has been pointed out, it can be corrected by simply inverting bit 6. An all-correct transmission would have produced an all-zero syndrome.

Table 2.8   Using Hamming codes, and finding a syndrome

Bit Position	1	2	3	4	5	6	7
(Binary)	001	010	011	100	101	110	111
	(P)	(P)	(M)	(P)	(M)	(M)	(M)
Check 1	*		*		*		*
Check 2		*	*			*	*
Check 3				*	*	*	*
Message			0		0	1	1
Parity bits	1	0		0			
Transmitted Code Word	1	0	0	0	0	1	1
Received Code Word	1	0	0	0	0	0	1
Recheck of Parity(reverse order)	0	1		1			

Syndrome = 110 = 6

Since there is an error in position 6, this can be simply corrected by inversion.

By adding an overall parity (O/P) check bit as shown in Figure 2.12(b), the single error correction capability is extended to double error detection. The error patterns indicating the following conditions:

1. No errors; zero syndrome and overall parity satisfied,
2. Single correctable error; non-zero syndrome and overall parity fails,
3. Double errors, non-correctable; non-zero syndrome and overall parity satisfied.

Hamming code schemes are particularly useful for systems that require a high level of data integrity and operate in noisy environments that create random bit errors.

Cyclic redundancy check (CRC)

This error detection concept is most effective in combating burst errors. To generate the code for transmission, three code words are used. A message code word k, a generator code word G (selected to produce the desired characteristics of blocklength and error detection/correction capability), and a parity check code word c. As for block codes, the transmission code word length is given by $n = k + c$. During encoding, the message k is loaded into a shift register and then moved c bits to the left, to make room for c parity bits. The register contents are then divided by the generator code word to produce a remainder that forms the parity check code word c, which is then loaded into the remaining shift register cells.

Thus if the total code word were transmitted and received without error, this when divided by G, would yield a zero remainder. The last c bits can then be discarded to leave the original message code word. If however, an error occurs, then division by G leaves a remainder code word that acts as a syndrome. There is a one-to-one relationship between this and the error pattern, so that any correctable errors can be inverted by the error correcting logic within the decoder. The effectiveness of these codes depends largely on the generator code word which has to be carefully selected. A further advantage is that CRC can be operated with microprocessor based coding, when the system characteristics become reprogrammable.

Other code error control techniques

There are a number of other techniques that have been developed from the basic Hamming concepts to deal with both random and bursts of errors. These include BCH codes for random error control, Golay codes for random and burst error control and Reed-Solomon codes for random and very long burst errors with an economy of parity bits.

Codes and coding formats

With all forms of signal transmission, it is important to minimise the effects of noise which is a destroyer of information. With analogue systems, the effect is quantified by the signal to noise (S/N) voltage, current or power ratio. In digital systems, where the information is transmitted in binary digits or bits, a similar concept can be applied. In this case the parameter is usually measured in terms of energy/bit per watt of noise power (E/B). If the noise power becomes comparable with the energy in each bit, bit errors are produced. Thus the degradation of the digital S/N ratio leads to a bit error rate (BER).

Primary codes and pulse shapes

The bit error rate can be minimised by using pulses of maximum width and/or amplitude, the obvious choice being a square shape. However this introduces a number of problems. To pass a square wave, a transmission channel requires a wide bandwidth. To retain a good approximation to a square wave requires that the channel bandwidth should extend up to at least the 13th harmonic of the fundamental frequency. In any case the transmission of such pulses through a typical channel will produce dispersion or pulse spreading, which leads to inter-symbol-interference and an increase in the bit error rate. Increased pulse width reduces the signalling speed and an increase in pulse amplitude introduces further problems. The final pulse shape is thus a compromise. One particular pulse shape is described as a raised cosine. This is chosen because half of the pulse energy is contained within a bandwidth of half the bit rate.

Secondary codes and formats

A code format is an unambiguous set of rules that defines the way in which binary digits can be used to represent alphabetic, numeric, graphic and control character symbols.

Shannon's rule states that the communication channel capacity in bits/second, is related to the available bandwidth and the S/N ratio. It also shows that these two parameters can be balanced to maximise the channel capacity for an acceptable bit error rate suitable for a particular service.

To take advantage of this trade off, binary code formats are designed by inserting extra bits into the data stream in a controlled way. Some of the formats used are shown in Figure 2.13. The general aim is to minimise the number of similar consecutive bits and balance the number of 1s and 0s in the message stream. The greater number of signal transitions is used to improve the locking of the receiver clock and so reduce bit errors. The balance between the number of 1s and 0s produces a signal without a dc component in its power spectrum. This allows ac coupling to be used in the receiver and reduce its low frequency response requirement. The commonly used codes are generated and decoded using dedicated integrated circuits (ICs).

A non-return-to-zero (NRZ) basic code is shown in Figure 2.13(a), where a 1 is signified by a full width pulse and a 0 by no pulse. A further variant of this is shown in Figure 2.13(b), where a 1 is signified by a signal transition at the bit cell centre and a 0 by no transition. Further variants of the NRZ code which have the same characteristics include inversions of these two. The return-to-zero (RZ) format where a 1 is represented by a half width pulse and a 0 by a negative pulse finds little use because the reduced width pulses represents an energy/bit penalty.

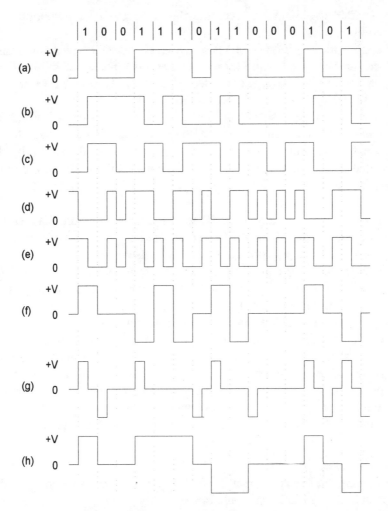

Figure 2.13 Commonly used code formats: (a) non-return to zero, (b) NRZ-M version, (c) Miller, (d) code mark inversion, (e) Manchester (bi-phase), (f) alternate mark inversion, (g) dicode, (h) duo-binary

Miller code is a popular format shown in Figure 2.13(c) and often used with magnetic storage media. A 1 in the original format is represented by a transition at each bit cell centre and a 0 by no transition, with the following exception: after two consecutive zeros an additional transition is introduced at the end of the first zero cell. A variant of this format utilises a transition for each 0 and no transition for a 1 with the opposite exception. In this format, a dc component is produced whenever an even number of 1s occur between two 0s. The Miller[2] format which

41

is very similar, avoids this by omitting the last previous 1 transition, whenever this condition arises.

The code mark inversion (CMI) is a two level code where a signal transition is introduced at the bit cell centre for a 0, so that 0 is represented by 01. There is no transition for a 1 which is represented alternately by 00 and 11 as shown in Figure 2.13(d).

The Manchester code shown in Figure 2.13(e) is a bi-phase format in which each bit in the original signal is represented by two bits in the derived format. The basic rule for the transform is that 0 is represented by 01 and 1 by 10. This ensures that there is never more than two identical bits in series. One variant of this format uses the opposite of this transform whilst two further variants adopt their inverses.

The alternate mark inversion (AMI) format shown in Figure 2.13(f) is a three level or ternary code where a 0 is represented by no transition whilst a 1 is represented by alternate positive and negative pulses.

The dicode shown in Figure 2.13(g) is another ternary format. The half width pulses are formed by differentiation of the original signal. Thus positive and negative signal transitions generate positive or negative pulses in the final format but with the same energy/bit penalty as return-to-zero codes.

The duo-binary ternary format has several variants, one of which is shown in Figure 2.13(h). This is generated from the original NRZ format by the transform, 0 is represented by no transition and a 1 by alternate positive and negative pulses according to the rule; Positive if the 1 is preceded by an even number of 0s and negative if the 1 is preceded by an odd number 0s. On average this new code has no dc component and a bit rate that is half that of the original signal.

Test Questions

1. Convert the binary number 0110,1001 into denary assuming that the code is in both 8421 BCD and pure binary.
2. Convert the following numbers into denary assuming that the two binary codes 1000,1111 and 0101,0000 represent signed numbers.
3. Convert 576(10) into both hexadecimal and BCD codes.
4. Using the complements method, perform the following subtraction: F002(H) – 10A3H.

3 Shift registers, counters, encoders/decoders, and multiplexers/demultiplexers

Syllabus references: D0–2.1, D02–2.2, M01–1.3, M01–1.5.

NOTE: *Before* reading this chapter, it may be advisable to revise Chapter 9 of Volume 2, Part 1, of this series which contains much of the basic information regarding these topics.

Shift registers

These devices are classed as sequential logic elements with outputs that depend not only on the current input, but in many cases, also the previous outputs. This feature allows them to act as limited capacity memory elements or data buffers. However, shift registers, of which there are three basic types, have a very much wider range of digital applications than this simple explanation suggests. As explained earlier, the R-S type suffers from an indeterminacy of output when $R = S = 1$. The D-type device which contains internal steering logic, avoids this problem by being effectively driven from a single input. The distinction between a D-type flip-flop and a data latch is small; the latch output follows the data input as long as the clock pulse is High, whilst the flip-flop data input is only transferred to the output during a clock pulse edge. By convention, the flip-flop is said to be Set when $Q = 1$ and Reset when $Q = 0$.

43

Digital Techniques and Microprocessor Systems

Figure 3.1 shows a hard wired version of a 4-bit shift register based on D-type flip-flops with AND-OR control logic. This provides for parallel data loading with serial format output and, in addition, parallel access to the same data stream.

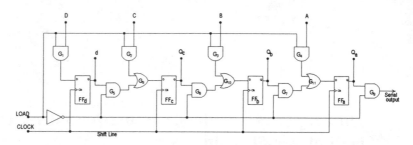

Figure 3.1 Parallel-in, serial-out (PISO) shift register with parallel access

With the Load line set to logic 1, gates G_1, G_2, G_3 and G_4 are enabled and any input at logic 1 is loaded into the D inputs. Similarly, any inputs set to logic 0 will simply provide a zero input. At the same time, gates G_5, G_6, G_7 and G_8 will be disabled to prevent any data movements. When the Load line is set to logic 0, the serial gates are enabled and the load gates disabled. The data bits can now be transferred to the Q outputs on the negative going edge of each succeeding clock pulse. After 4 clock pulses the shift register is again ready to be loaded with the next nibble.

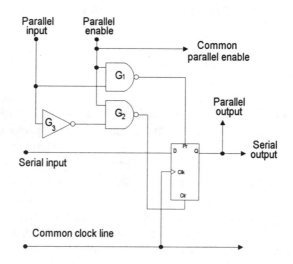

Figure 3.2 Alternative parallel load shift register

44

Figure 3.2 shows an alternative approach to achieve parallel loading of a D-type shift register. This uses the input logic level to force the Q output to take up the same state via the preset and clear inputs. When the parallel Load line and the data input is High, Pr is driven Low and Clr High through the steering logic, to force the Q output to the logic 1 state. If the data input is at logic 0, the states of Pr and Clr are reversed and the Q output is forced to logic 0. Once loaded, the parallel load line is disabled and the register data is right shifted on the negative going edge of each clock pulse.

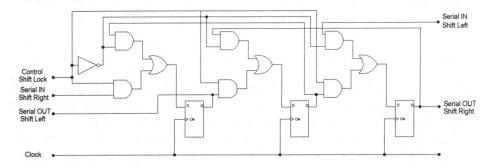

Figure 3.3   3 stages of an 8-stage shift left / shift right register

Figure 3.3 shows a simplified diagram of an 8-bit dual purpose shift register. In the practical device, the central flip-flop and its control logic is duplicated as often as needed. With the shift control line set to logic 1, all the lower AND gates are enabled so that data can be right shifted at each clock pulse. Conversely, with the control bit set to 0, the upper AND gates are enabled so that data can be left shifted.

The J-K flip-flop, which also contains steering logic to avoid the ambiguity of output, also has a greater range of output controls as shown by Table 3.1.

Table 3.1   State table for a J-K flip-flop

J	K	Q_N	Q_{N+1}	Comment
0	0	0	0	No change
0	0	1	1	No change
0	1	0	0	Reset to 0
0	1	1	0	Reset to 0
1	0	0	1	Set to 1
1	0	1	1	Set to 1
1	1	0	1	Toggle action
1	1	1	0	Toggle action

(Q_N = state now; Q_{N+1} = state after next clock pulse)

45

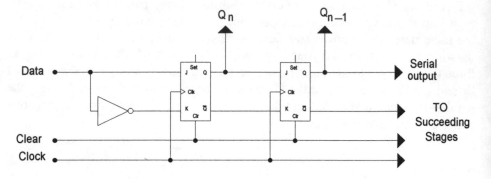

Figure 3.4   J-K shift register with parallel access

Figure 3.4 shows two sections of a J-K shift register for serial output but with parallel access. As can be seen from Table 3.1, the J and K inputs not only control the flip-flop action, but also provide the Set and Reset functions.

If the complementary outputs of the J-K device change state before the end of a clock pulse, then because of the internal feedback, the inputs will also change. This can cause the device to oscillate until the end of the pulse and leave the output in an indeterminate state.

To avoid this so-called *race around*, the clock pulse duration should be small compared with the propagation delay. For high speed operation, this is avoided by using a master/slave device as shown in Figure 3.5 where the NAND logic gates act as switches.

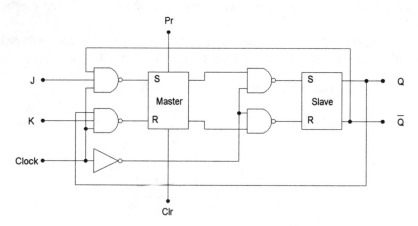

Figure 3.5   Master/slave J-K flip-flop

Data is transferred in two stages on both edges of the clock pulse. On the positive edge, the input gates are operative and allow the data to be loaded into the master flip-flop. At this time the output logic gates are open circuit. On the negative pulse edge, the switch states reverse, the master is isolated from the input and the data is transferred to the slave stage to provide the output.

Counters

As shown earlier, flip-flops are basically divide-by-two devices that may be clock driven either in the serial or parallel mode. The output from a string of series coupled flip-flops thus produce a cyclical bit pattern that is binary weighted. The all 1s count is followed by all 0s to restart the sequence. In a complementary manner, if the clock frequency is very accurately controlled, it becomes possible to convert the count into a time duration. Hence the development of measuring instruments known as counter/timers.

Modulo-N counters

The maximum count value is used to define the modulus of the device. For example, a 3-bit counter has a maximum range of $2^3 = 8$ and is described as a *modulo-8* counter. Similarly, an instrument constructed from 8 flip-flops would be described as modulo-256. Actually the counter output value is always the remainder of the total count after division by the modulus.

It is possible to construct a counting instrument that operates to a modulus other than a power of 2, hence the term modulo-N counters. To construct such a counter, start with a chain of n flip-flops such that n is the smallest number for which $2^n > N$. Add a feedback loop and NAND gate such that on the count of N, all the stages are reset to zero. Each NAND gate input is the set of Q outputs that are set to 1 on the count of N. The gate output then sets all flip-flops to zero via the parallel connected clear inputs. For example, a decade counter requires 4 flip-flops because $2^4 > 10$. Since $10 = 1010$ binary, bits 2 and 4 are set to 1 on a count of 10 and these provide the NAND gate inputs.

Figure 3.6 shows the construction of a simple modulo-5 counter. Due to the propagation delay in a long chain of flip-flops, the clear operation may not occur cleanly. The first element may clear before later ones so that a narrow extra pulse output is generated. This can be suppressed by the addition of a latch circuit in the feedback loop.

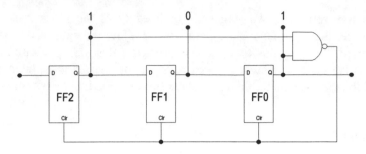

Figure 3.6   Modulo-5 counter control logic

Exercise 3.1

Using a proprietary logic trainer board; design, construct and test a modulo-7 counter circuit.

Figure 3.7 shows the basic principles of a counting device that has been adapted to measure frequency. A crystal controlled oscillator frequency is divided down to produce an accurate time gating signal. In practice, the frequency is chosen to be high enough to provide the maximum count or minimum time ranges to meet the user demand. In the case shown, the basic frequency is 1MHz which when divided by 10^6 provides a time gate of 1 second. The input signal is converted into a square wave by suitable circuit and then input to the AND gate which is enabled by the timing gate signal. The displayed count then represents the unknown frequency in cycles per second or Hz.

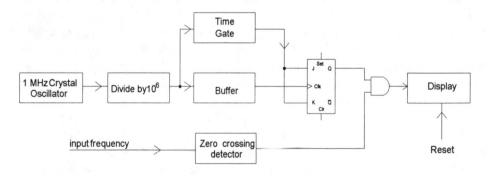

Figure 3.7   Frequency measuring counter

Figure 3.8 shows how the same basic components can be organised to produce a timer circuit. The crystal controlled oscillator output is again divided down to produce a suitable enabling signal to the AND gate which is further controlled via

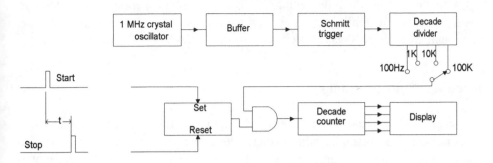

Figure 3.8   Time interval measurement by pulse counter

a Set/Reset (S/R) bistable device. When Set and Reset by start and stop signals, the display output represents the time delay t, between the two pulses. For example, if the decade divider output of 100kHz is selected and a count of 87 is measured, then the time $t = 10^{-5} \times 87 = 0.87$ms.

Encoder/decoders and multiplexers/demultiplexers

The basic principles of these devices were discussed in Chapter 9, Volume 2, Part 1, of this series. The reader is thus referred to this source for any necessary background reading.

Since these concepts have much in common with each other, many ICs are available that are capable of being used in either application. The similarities can be clearly seen from the following summary.

Encoder. A device with N inputs, only one which can take up the logic 1 state to generate a unique N-bit parallel output code.

Decoder. A device for which an N-bit input word establishes a 1 state on one and only one, of 2^N output lines.

Multiplexer (Mux). Also known as a *data selector.* A device that selects 1 out of N input data sources to be transmitted over a single output line.

Demultiplexer (Demux). Transfers an input single serial binary data stream to a selected 1 of 2^N output lines.

Because many microprocessor controlled systems are now capable of handling both analogue and digital signals, it is important to recognise that multiplexer ICs are available that can be used to distribute signals in either format. An important parameter of these devices is the settling time (T_s) which relates to the maximum switching rate between channels. Due to the device self capacitance, the response to a step input becomes slewed and limits the maximum change over rate. The settling time is therefore defined as the time lapse between the onset of the step input, to the time at which the output settles to a level close enough to the ideal to ensure a secure switching action. T_s is typically less than about 4µs, so that the maximum switching rate is the reciprocal of this value, about 250kHz.

Encoders and decoders

An encoder is necessary when a particular input function must generate a unique code pattern. For example, when using a hexadecimal keypad, each of the 16 input characters must generate a unique 4-bit code. Thus such an encoder would require 16 inputs, only one of which is selected at any one time, and 4 output lines to provide the parallel bit pattern. In a complementary way, the decoder would have 4 input lines in order to select one and only one of 16 output characters.

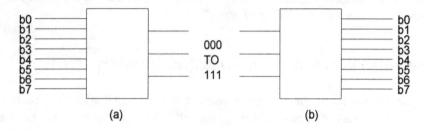

Figure 3.9 **(a) 8 to 3-line encoder, (b) 3 to 8-line decoder**

Figure 3.9 shows the basic principle of this technique for a simple 8 to 3 line encoder and the complementary 3 to 8-line decoder. It may be necessary to operate the decoder at specific times and because the bit patterns are in a parallel format, it is important to maintain synchronism.

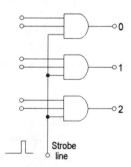

Figure 3.10 **Strobing decoders**

50

Figure 3.10 shows a common way of achieving this through the use of a single synchronising or strobe pulse. The bit pattern present at the input will cause the correct output code to be generated at the time of the strobe pulse.

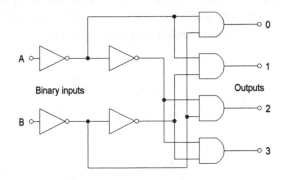

Figure 3.11   2-line to 4-line decoder

Figure 3.11 shows the logic circuit of a 2 to 4-line decoder designed to decode the 2-bit input pattern A,B and generate a logic 1 to select a particular output line or bit pattern.

Table 3.2 provides the truth table for this device from which it can be seen that for an input of 0,0 the output pattern will be 1,0,0,0, showing that only line zero has been selected. It is left as an exercise for the reader to confirm the validity of the other output values.

Table 3.2   Truth table for the decoder

Inputs		Outputs			
A	B	0	1	2	3
0	0	1	0	0	0
0	1	0	1	0	0
1	0	0	0	1	0
1	1	0	0	0	1

Exercise 3.2

Confirm the validity of the truth table given in Table 3.2 above for the 2 to 4-line decoder shown in Figure 3.11.

Multiplexers and demultiplexers

The logic circuit of a device designed to transfer the data on one line of input to a single output channel is shown in Figure 3.12. The particular input line is selected by the 2-bit input, referred to as the channel address. Data from an input line is transferred to the transmission channel only during the period for which the strobe line is active low.

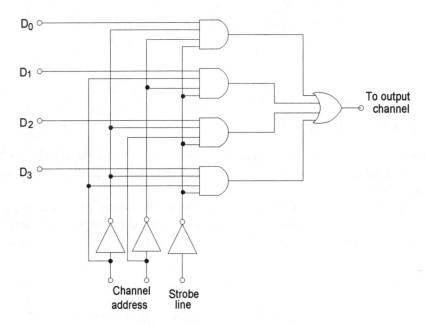

Figure 3.12   4 to 1-line multiplexer

If the strobe line is driven high, the multiplexer is disabled and this feature is used in Figure 3.13 to enable two 4 to 1-line multiplexers to be expanded for 8 channel operation. In both circuits, the channel addresses may be generated by a counter as shown in Figure 3.13 or from a programmable sequence generator.

The circuit shown in Figure 3.14 is that of a 1 to 4-line demultiplexer/decoder. When the strobe line is active low, the time multiplexed input data is transferred to the output channel selected by the two bits A and B in the manner shown in the Table 3.3.

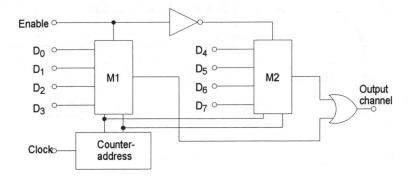

Figure 3.13   Multiplexer extension

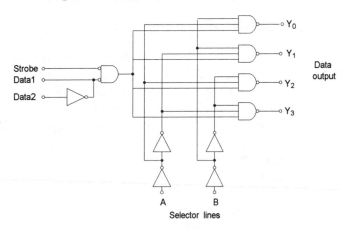

Figure 3.14   1 to 4-line demultiplexer / decoder

Table 3.3   Multiplexing action

Address		Data	Strobe	Outputs			
A	B			Y_0	Y_1	Y_2	Y_3
0	0	0	0	0	1	1	1
0	1	0	0	1	0	1	1
1	0	0	0	1	1	0	1
1	1	0	0	1	1	1	0

If the decoder control line is set to logic 1, this circuit will behave as a 2 to 4-line decoder, the two input lines being the selector lines A and B. The selected output line is driven low at the time of the strobe pulse. The operation is explained in Table 3.4.

Table 3.4   2 to 4-line decoder action

Select		Control	Strobe	Outputs			
A	B			Y_0	Y_1	Y_2	Y_3
0	0	1	0	0	1	1	1
0	1	1	0	1	0	1	1
1	0	1	0	1	1	0	1
1	1	1	0	1	1	1	0

Exercise 3.3

Using a logic trainer or tutor, investigate the actions of the multiplexer/de-multiplexer and encoder/decoder devices described in this chapter.

Test Questions (drawings follow)

1.   Figure 3.15 shows two 7474 ICs. Complete the diagram to produce a 4-bit serial input, parallel output shift register. Label all the inputs and outputs.
2.   Draw the truth table for the logic diagram shown in Figure 3.16. Describe the circuit operation and state a function that it could perform.
3.   A circuit is to be designed to produce an output of logic 1 if there are two or more 1s in a 3-bit input code. Draw the truth table to deduce the Boolean expression for this function. Using Karnaugh mapping reduce this to its simplest case and then draw the logic diagram.
4.   Draw the truth table for the device shown in Figure 3.17. Explain how the device may be used as an address decoder to allocate addresses from 0000H to 3FFFH for four 4k EPROMs.

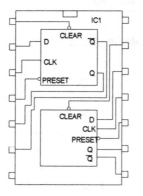 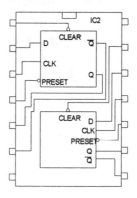

Figure 3.15   Figure for Question 1

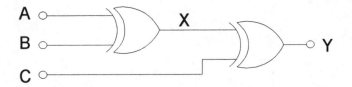

Figure 3.16   Figure for Question 2

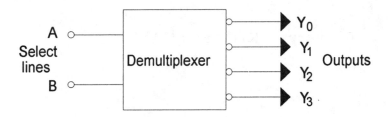

Figure 3.17   Figure for Question 4

4   Memory — ROM and RAM

Syllabus references: D03–3.1, I2–2.3, M03–3.1, M06–6.1

NOTE: You should be familiar with the basics of memory as discussed in Volume 2, Part 1, Chapter 10, of this series.

Addressing

Memory is a name for a type of circuit component which can be of several different types. The basis of a single unit of memory is that it should retain a 0 or a 1 level, and that this voltage can be connected to external lines when needed. The method that is used to enable connection is called *addressing*. The principle is that each unit of memory should be enabled with a unique combination of signals that is present on a set of lines, called the address lines or the address bus, Figure 4.1.

The name 'address' comes from the comparison with posting letters to house addresses. Since each combination of bits can be represented by a number, the combinations are called address numbers. Since these numbers are long and clumsy when written in binary, it is usual to express address numbers in hexadecimal — consult Volume 2, Part 1, if you are not familiar with the hex scale of numbers.

If we stay for the moment with the idea that memory consists of stores of voltage levels that represent binary digits, and that these are connected to external

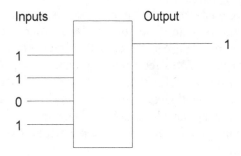

Input is 1101B = 0DH = 13 denary
Output is 1 – the value stored at address 13

Figure 4.1 A simple memory chip showing how an address number is applied and causes a stored digit to become available

pins by using an address number, then we can examine the methods that are used to implement this principle.

To start with, there are two basic types of memory, both of which are needed in virtually any microprocessor application. One type of memory consists of fixed bits, unalterable, and therefore called 'Read-only' memory, or *ROM*. The important feature of ROM is that it is *non-volatile*, meaning that the stored bits are unaffected by switching off power to the memory, and are available for use whenever power is restored. Since there must be an input to the microprocessor whenever it is switched on, ROM is essential to any microprocessor application, and in some applications it might be the only type of memory that is needed.

The simplest type of such a ROM consists of permanent connections to logic 0 or logic 1 voltage lines, as Figure 4.2 indicates. In this diagram, two AND gates are shown, with one input of each tied to a fixed voltage.

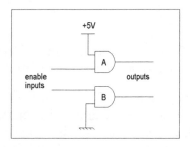

Figure 4.2 How masked ROM is constructed, using gate connections to fixed logic levels

When the gates are enabled by the enable lines rising to logic 1, gate A output will be at logic 1, and the gate B output will be at logic 0 because of the fixed connections at the inputs. We can develop this idea into the system shown in Figure 4.3, in which eight gates feed into a common output line. One input of each gate is held at a fixed level, either 0 or 1, and the other input is used as an enable, fed from a demultiplexer. In this way, the three lines into the demultiplexer carry binary signals which will activate any of eight lines out. These output lines will each in turn activate one gate, so passing one signal, 0 or 1 to the single output.

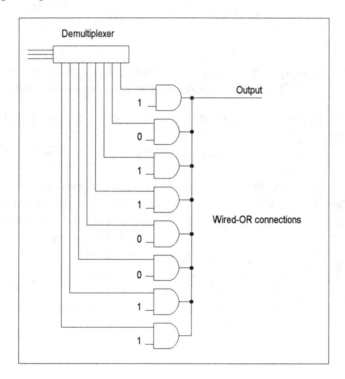

Figure 4.3   A simple address circuit, allowing address signals on three lines to activate fixed signals from the output

The remarkable developments in IC technology now allow economical and comparatively straightforward manufacturing of circuits of this type which have eight sets of demultiplexers and gates, each using 16 address lines and with 65,536 gates to each output pin. In other words, this is a 64K x 8 bit ROM. This type of ROM is called a 'masked' ROM, referring to the IC manufacturing technique in which etching masks determine the layout of connections. The masks form the main initial cost of production of such a ROM, and the use of masked

ROM is feasible only if the content of this memory is thoroughly tested and proven.

For example, the IBM PC type of computer uses one masked ROM to store the elementary program codes (the BIOS) that are essential to make it possible to use the disk drive and which provide access for other control actions. This ROM runs the machine when it is first switched on, and its main action is to read the rest of the operating system from a hard disk, after which the operating system takes over, using the ROM only for access to the microprocessor.

An alternative to masked ROM is some form of EPROM, the electrically programmable read-only memory. There are several varieties, but nearly all use the same principles of making connections through lightly doped semiconductor by injecting carriers (electrons or holes), which are then trapped. The system of gating and demultiplexing which forms the addressing for the chips is the same, only the method of connecting to logic 1 or 0 through paths in the semiconductor is different.

The point about a PROM is that the connections can be established by connecting to the inputs of gates like the gates of Figure 4.3. These are then 'blown' by using higher than normal voltages (typically 16 to 25V for a chip that normally operates at 5V), in a programming cycle. This normally consists of cycling several times through each address number, applying the high voltage for each logic 1 bit that is needed, with the correct signals taken from a temporary memory source.

Once programmed or blown in this way, the PROM can be used like a ROM. The advantage is that a PROM is manufactured blank, there is no special masking cost, and, more importantly, no extra design time needed in the manufacturing process. If the programming is faulty, new PROMs can be blown and tried, until the system seems to be trouble-free. If this is done at prototype stage, the result can be a very reliable piece of equipment, and a masked ROM can be made from any copy of the PROM.

Using PROMs is a short-term expedient, but one which is useful in that short term. Though PROM chips are expensive, they can be re-used, apart from the 'fusible link' type in which the internal connections are opened permanently during the blowing process. The most popular type of PROM, the EPROM, is erasable by shining ultra-violet (UV) light into the silicon whose conductivity establishes the logic 1 connections. The effect of UV is to make the semiconductor material conductive to such an extent that the trapped charges can move out, making the material into an insulator. Figure 4.4 illustrates the pinout for a typical EPROM, and the symbol that is used to indicate this type of component.

Digital Techniques and Microprocessor Systems

Using the example, programming is carried out with each address established on the address bus in turn. For each address, the data is placed on the data lines, and the PD/PGM pin is pulsed to place the data into the EPROM

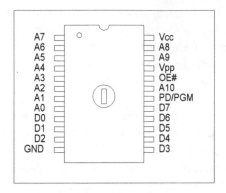

Figure 4.4   A typical EPROM symbol and pinout diagram

The process of erasure is described as 'washing', and typically takes an exposure of 5 minutes to 30 minutes, depending on the construction of the PROM and the wavelength of UV. Exposure is carried out by shining UV through the small quartz 'window' over the chip; this is the meaning of the symbol at the centre of the EPROM chip symbol. The most effective UV for the purpose is the shorter wavelength type, and this radiation must not be allowed to reach any part of a human body, particularly the eyes. PROM-washers must therefore be constructed in light-tight boxes, with interlock switches to eliminate the possibility of the light being on when the box is open.

Another form of PROM is the *EAROM*, the electrically-alterable ROM, in which internal links can be made by using higher than normal voltages on the data lines when the memory is written. These links are then unaffected by normal operating voltages, but the whole memory can be re-programmed by using the higher voltages when required. EAROM has been used in programmable calculators, but it is not found to any extent in computers because of slow response times.

One point to note is that PROM and to a lesser extent, ROM, is slow-acting. If a computer needs to make frequent reference to codes contained in ROM or PROM, and if speed is important, as it will be in a machine using a clock speed of 50MHz or more, then the ROM/PROM codes can be copied into fast RAM. This is known as 'shadowing', and its use is now very common for modern fast computers.

Read-write memory (RAM)

Read-write memory is the other form of memory which for historical reasons is always known as RAM (Random access memory). This is because in the early days, the easiest type of read-write memory to manufacture consisted of a set of serial registers, from which bits could be read at each clock pulse. The construction of a memory from which a bit could be selected by using an address number was a much more difficult task. The use of addressing means that any bit can be selected at random, without having to feed out all the preceding bits, hence the name 'random access'. Practically all forms of memory that are used nowadays in microprocessor systems feature random-access, but the name has stuck as a term for read-write memory.

Unlike ROM, RAM can be *volatile* or *non-volatile*. The earliest generations of mainframe computers used RAM which was constructed from magnetic cores. Like any storage scheme based on magnetisation of a material, this was a non-volatile memory system, so that data remained in the RAM even after the machine was switched off. Modern RAM systems use IC chips, and completely non-volatile RAM is not possible. It is possible, however, to fabricate memory using CMOS techniques, and retain data for very long periods, particularly if a low-voltage backup battery can be used. Such CMOS RAM is used extensively in calculators, and it is also used in desktop computers, with battery backup, as a way of holding information (about disk drives, for example). Failure of a desktop machine to switch on correctly is often caused by battery failure, though the modern lithium batteries which are used in recent designs are likely to outlive the rest of the machine, since most desktop machines become out of date in three years or less. For the most part, however, RAM is volatile, and the main divisions are into static and dynamic forms.

Static RAM is based on using a flip-flop as each storage bit element. The state of a flip-flop can remain unaltered until it is deliberately changed, or until power is switched off, and this made static RAM the first choice for manufacturers in the early days. Figure 4.5 shows the most elementary type of flip-flop using only two bipolar transistors — MOS transistors are normally used.

Static memory will, in practice, use flip-flop units that contain more than two active elements, so that faster switching can be achieved. This means that power consumption will be comparatively large, because each flip-flop will draw current whether it stores a 0 or a 1. This has led to static RAM, except for the CMOS variety, being used only for comparatively small memory sizes. The resistor loads illustrated in Figure 4.5 would, in practice, be transistors.

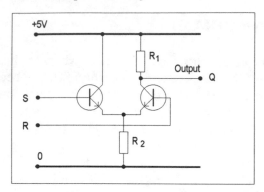

Figure 4.5 **The basis of static RAM is the flip-flop. When one transistor passes current, the output state is logic zero; when the flip-flop changes over the output state is logic 1. The current drain is the same in each case**

The predominant type of RAM technology for large memory sizes is the dynamic RAM. Each cell in this type of RAM consists of a miniature MOS capacitor. Logic 0 is represented by a discharged capacitor, logic 1 by a charged capacitor. Since the element can be very small, it is possible to construct very large RAM memory chips (1M x 1 bit, 4M x 1 bit are now quite common), and the power requirements of the capacitor are very small. The snag is that a small MOS capacitor will not retain charge for much longer than a few milliseconds, since the connections to the capacitor will inevitably leak. All dynamic memory chips must therefore be *refreshed*, meaning that each address which contains a logic 1 must be re-charged at intervals of no more than a few milliseconds.

The refreshing action can be carried out within the chip, providing that the cycling of address numbers is done externally. Some microprocessors (notably the old Z80) provide for this internally, others require the use of additional refresh circuitry. The availability of very large capacity dynamic memory chips has caused the price of RAM (expressed as pence per kilobyte) to fall dramatically for a considerable period. Over the same period, the reliability of dynamic RAM, which at one time was suspect, has improved so as to be on a par with any other IC components.

Because of the need for refresh signals, dynamic RAM is addressed in a way that differs from the methods used for static RAM. Static RAM uses a set of address pins and a data pin, along with a read/write pin. To write data, the address is placed on the address lines as a binary number, the data is placed on the data line, and the write pin is activated (usually by taking it to logic 0). To read the data, the address is again established, and the read pin is activated so that the data bit is available on the data line.

Dynamic RAM uses a smaller set of address pins than its static counterpart, along with two pins that are marked as CAS and RAS, meaning Column Address Strobe and Row Address Strobe. A single memory cell is located in two steps, first by activating the CAS pin and placing a column address on the address lines, then by activating the RAS pin and placing a row address on the address pins. Figure 4.6 illustrates a Toshiba chip (TC511000P-10) which uses ten address pins labelled in the usual way as A_0 to A_9. Ten digits of binary allow for a count of 1,024, so that 1,024 column numbers and 1,024 row numbers can be used, giving a total of 1,024 x 1,024 addresses, which is 1,048,576 bits; 1Mb of RAM.

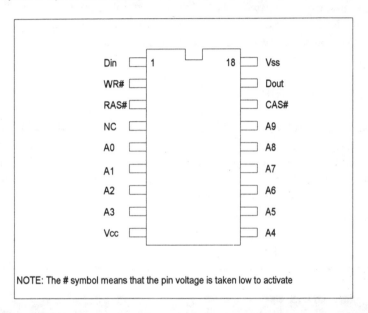

NOTE: The # symbol means that the pin voltage is taken low to activate

Figure 4.6   A typical dynamic RAM chip pinout. This example is a 1Mb chip

By using the CAS and RAS pins in conjunction with the address pins at times when the microprocessor does not use memory, the memory cells can be refreshed — typically at the rate of 512 refresh cycles each 8ms. Standby current is 2mA, and maximum current (with all cells holding logic 1) is 60mA. The larger current is needed because of the current required to refresh the large number of capacitors at the rate of 512 in each 8ms.

The requirement for RAS and CAS signals, along with refreshing, means that a dynamic RAM controller chip (such as the 74S409) must be used unless the microprocessor provides for refreshing. For industrial controllers, static RAM is more usual, so that no refresh problems arise. One microprocessor chip, the Z80, incorporates its own refresh capability, so that it can use dynamic RAM with the minimum of other circuitry.

It is this development in dynamic RAM technology which more than any other single factor, has been responsible for the simultaneous drop in prices and rise in memory size of small computers over the past ten years. More recently, improvements in CMOS technology have resulted in large sizes of static RAM chips which need no refreshing. The problem of early CMOS designs — slow operation — has been overcome, and the slightly higher cost of modern static RAM is balanced by the lack of any need for refresh circuitry. Nevertheless, the very large memory sizes (typically 8Mb) of modern computers require the use of dynamic RAM for all except fast cache memory (used for temporary storage and capable of operating at the speed of the microprocessor).

Memory requirements

The most obvious requirement for memory is its size, and this will be the first consideration for selection of memory. Modern desktop computers do not use memory units in single-chip form, and the normal method is to connect up a set of chips into a unit called SIMM, single inline memory module. The early type of SIMM used 36-pin strip construction (Figure 4.7).

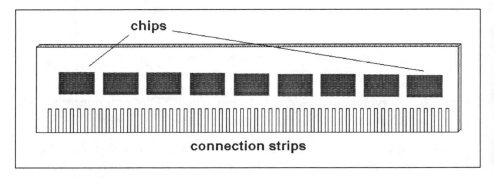

Figure 4.7   A typical older type of SIMM card as used on older PC computers. Modern units use 72 pins and eight or nine memory chips, with memory arranged in four banks

Modern SIMM units use a 72-pin construction, and are arranged so that the memory chips are organised in 36-bit sets rather than the older 8-bit grouping. This allows memory to be 'banked', a system that allows the computer to operate faster than the response time of the memory by using different parts of the memory for consecutive reads. For example if a 4-byte unit is to be read it would not be read from one memory chip, but from four chips activated in turn. This allows each chip to recover before it is used again, and also allows time for refreshing dynamic RAM.

Note that self-refreshing memory chips are now available. These are used for specialised purposes, such as memory cards for portable computers.

Memory speed

There are two times involved in using memory. The *access time* is the time that is needed between placing an address on the address lines, and the data bit(s) becoming available on the data line(s). This applies for both reading and writing, because if a bit is placed by the microprocessor on to the data line before the address has become established, the bit will not be stored or, worse, will be stored in the wrong location because some addresses may be falsely activated until the address lines settle.

The other important time is the time needed to read or write the data once it is available. This depends on the microprocessor as well as on the memory chips, but because modern microprocessor chips can operate at very high clock speeds, faster than dynamic memory can cope with, it is the response time of the memory that is the limiting factor.

A total response time of 70ns is nowadays considered to be a minimum requirement for desktop computers, and some machines demand lower times. For less stringent requirements, 100ns or more may be acceptable, and the longer access times make chip design easier, leading to lower prices.

Even a response time of 70ns, however, is inadequate for computers whose clock rate may be 50MHz or more. At 50 MHz, the time between clock pulses is 20ns, considerably less than the memory access time, and when a computer works with 32-bit (4-byte) units, a memory access will require reading four bytes in as many clock pulses. This will be followed by several clock cycles in which no access will be required, but the problem is that the address numbers have to be established before the data can be read or written, and the memory will still be active for some time after the read or write. This is dealt with firstly by *banking*, — storing the bytes in a set of four in different memory areas so that the first memory area can recover before another demand is made on it.

The second part of the solution is pipelining and caching. *Pipelining* uses a temporary (fast) memory store within the microprocessor, allowing addresses to be placed on the address lines before the data is needed — this requires the microprocessor to be designed so as to 'look ahead' for its addresses. *Cache* memory is also fast static memory that is filled from dynamic RAM and is used directly by the microprocessor. Since almost 90% of data is in a sequence, it is

90% certain that the microprocessor can read its next byte from the cache (or write its next byte to the cache). If the byte is not in the cache, it will have to be read from dynamic RAM, but because the dynamic memory is being used more efficiently this need not impose too large a time delay. The cache is filled from dynamic RAM, or written to dynamic RAM, at times when the microprocessor does not require memory access.

Using ROM and PROM

ROM and PROM can be used in controller applications for a variety of purposes other than the storage of program bytes. A typical application is as a character generator for displays.

Take, for example the usual seven-segment display as illustrated in Figure 4.8, which also shows the letter codes that need to be illuminated for displaying number from 0 to 9. As a logic problem, designing gates that will output seven-segment codes for inputs in the range 0 to 9 is by no means simple.

Number	Segments
0	abcdef
1	bc
2	abged
3	abgcd
4	fgbc
5	afgcd
6	afedcg
7	abc
8	abcdefg
9	abcdfg

Figure 4.8   **The arrangement of a seven-segment display, and the activated segments for numbers 0 to 9. Note that letters A to F can also be displayed, allowing hexadecimal displays**

Using a ROM is considerably easier, and Figure 4.9 shows how this could be organised. The inputs consist of four address lines (because four bits are needed for the number range 0 to 9, with several spare unused addresses), and the outputs are on seven lines that correspond to the segment letters. We can write the segment output as binary number, with 1 meaning a segment activated, so that

A	B	C		a	b	c	d	e	f	g
0	0	0	0	1	1	1	1	1	1	0
0	0	0	1	0	1	1	0	0	0	0
0	0	1	0	1	1	0	1	1	0	1
0	0	1	1	1	1	1	1	0	0	1
0	1	0	0	0	1	1	0	0	1	1
0	1	0	1	1	0	1	1	0	1	1
0	1	1	0	1	0	1	1	1	0	1
0	1	1	1	1	1	1	0	0	0	0
1	0	0	0	1	1	1	1	1	1	1
1	0	0	1	1	1	1	1	0	1	1

Figure 4.9 **Outline of a ROM that will permit the control of a seven-segment display for numbers 0 to 9 inclusive, and the table of inputs and outputs**

a zero would be represented as 1111110 (segments abcdef activated) and 2 as 1101101 (abdeg). The ROM can then be constructed so as to output the correct 7-bit codes for each four-bit input from 0000 to 1001.

A table of inputs and outputs can be used to program a ROM or PROM — if the outputs are 8-bit then the highest order bit is always 0. A PROM programmer would normally require the data to be in hexadecimal rather than binary.

Exercise 4.1

Using a PROM-burner and either a computer or a programmable microprocessor unit, create a PROM that can be used for driving a seven-segment display.

Character generators

Character generation for a VDU based on a cathode-ray tube (CRT) is more complex, and is easier to understand if a specific example is used. The screen mode that is commonly used on PC computers uses 25 lines or 80 characters per line, with each character taking up to 16 raster lines and up to eight dot spacings on a line.

Figure 4.10, overleaf, shows the word 'Bit' as it would be set out for a VDU display. One top space and two bottom spaces are normally left clear to allow for spacing between lines of characters, and only the characters with 'descenders', the lower portions of the letters g and y for example, use the lower spaces.

Figure 4.10   Using dots (pixels) to place a word on the screen

The problem is to see how this screen display of dots along lines can be produced from a ROM. If we take this example, the first scanning line for this set will be blank, with no output required. The second scanning line will require three dots for the top of the letter 'B', and the third will use three dots, differently positioned, for the second line of 'B' and the top of the letter 't', and so on for the whole set of 15 lines. Modern character generators deal with 8-bit codes rather than the 7-bit codes which were once used.

The storage needed to cope with all this consists of three parts. One part is a character store which must be RAM, since its contents will change for each screen 'page'. The second part is a ROM which allows the characters to be output as a set of dots, and the third part is a parallel-in, serial-out (PISO) register which allows the dot information to be fed out to the modulator of the CRT in sequence. Figure 4.11 shows this arrangement as a block diagram.

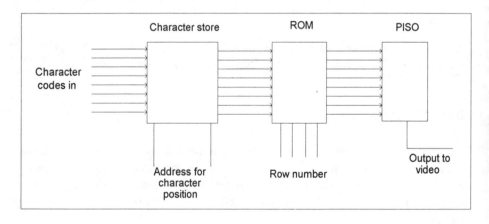

Figure 4.11   The arrangement of character store, ROM and PISO in a video character generator

The action is as follows. The character store is filled with the codes for a complete screen-full of characters. Since each line can contain 80 characters and there will be 25 such lines, this store must cope with 2,000 characters, each of 8 bits, so that the RAM requirement is 16,000 bits. To select each character will require an addressing input of 11 lines (because $2^{11} = 2,048$) so that each character can be selected both for input and for output.

With the character store filled, the ROM has its row number set for the first of the 15 lines that each character will use. The first character of the first line is read from the RAM store, and the ROM produces the bits that represent the top portion of the character (usually all blank). This is output to the PISO and to the video circuits and the second character is entered from the RAM so that its top row can also be analysed. When the 80th character has been read, all of the top row dots are on the screen, and the first scanning line of the set is complete.

On the second scanning line, the row number for the ROM is incremented, so as the same line of characters are read in, the ROM will produce the codes for the second row, output these to the PISO and so to the video circuits. This set of processes continues until all fifteen rows for a line of characters have been converted, and the next line of characters will also be read fifteen times from the RAM. When all 25 lines of characters have been scanned in this way, the whole process repeats from the start after the field interval of the display (in which the beam returns to the top left-hand corner of the CRT).

Number-base generators

Another use for ROM is in converting to other number bases, a modulo-N counter. The word modulo refers to the number base used for counting, and our familiar number-base is ten, signified by being written with a '1' in a tens column and a zero in the units column. You will by now also be familiar with modulo-2 or binary, and with modulo-16, or hexadecimal (hex).

As an example, consider a modulo-7 system, in which we count from 0 to 6, then the next output is 10 (denary 7), followed by 11 (denary 8) and so on to 20 which is denary 14 and so on. If we needed a modulo-7 counter there are two main ways of obtaining it. One is to use a register which can be connected to provide this count, and there are standard methods of designing such systems. The other is to use a ROM in which the input lines are used to produce modulo-11 outputs.

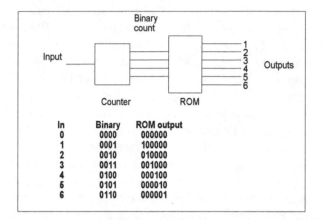

Figure 4.12   A modulo-7 counter using a binary counter and a ROM chip

Figure 4.12 shows an elementary counter of this type, which consists of a normal 4-bit binary counter coupled to a ROM in which six outputs are used. The programming of the ROM is also illustrated, so that you can see the outputs that correspond to the binary count inputs.

The output is the same as would be obtained from a Johnson counter constructed from flip-flops, using in this case each ROM output line for an individual digit. In this simple example, there is no provision shown for resetting the binary counter after a count of six, nor for carrying the modulo-7 count to another stage.

Sequencers

A sequencer is a circuit that produces a set sequence of outputs for a given set of inputs. A simple example is provided by traffic lights. The inputs are time signals which are set as short or long according to the density of traffic (using the sensor pads on the road).

The outputs are the lights sequence of Red — Red and Amber — Green — Amber — Red, and the problem is how to produce this sequence from a count of 0 to 3. This is easily dealt with using a binary counter and ROM (Figure 4.13, opposite).

There are three control outputs, one for each colour of light, and a single input for the timing pulses. The timing pulses are applied to a 2-bit binary counter, and these count numbers are applied in turn to the ROM, which produces the output sequence. The table in Figure 4.13 shows the ROM outputs as binary numbers in which the most significant figure switches the Red light and the least significant figure switches the Green light.

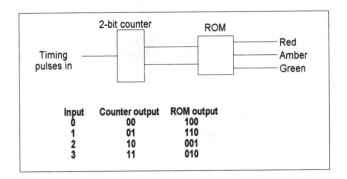

Figure 4.13   A traffic-light sequencer using a ROM

Sequencing is a very important action in machine control, and a simple example is an automatic washing machine. The input to a digital sequencer for a washing machine would consist of timed pulses, with each pulse input changing the binary input to a ROM. The outputs for the ROM would control the water inlet valve, the wash-action motor, the spin motor, the outlet valve, the water heater, and the water pump, and different wash cycles could be created by using different timings and different settings of the thermostat that controls water temperature.

Note that unless such a system is battery-backed, a mains failure will cause the system to reset, whereas a simple mechanical sequencer, when power is restored, would continue its action from where it was interrupted

More elaborate machine-tool controllers can use sequences of, for example, alternate measurement and cutting, shifting tool position, drilling, chamfering and finishing. Though there may be a very much greater number of steps, the basic principles remain the same, and the use of a counter and ROM can provide the necessary control. For the more elaborate controllers, the controller might use a PROM so that the settings could be changed when required, or the control codes might be read into a RAM from tape or disk as required.

Exercise 4.2

Burn a PROM so that it will produce the following set of outputs, given the number of pulses into a binary counter.

Pulse Number	Output	Pulse Number	Output	Pulse Number	Output
1	5	2	5	3	7
4	9	5	11	6	13
7	3	8	11	9	9

Exercise 4.3

Using RAM and a logic tutor, create a traffic-light sequence at the output. Could you alter the programming so that the red period could be made longer or shorter?

Exercise 4.4

Transform the table shown here into codes to be placed in EPROM. Your answer should be in the form of a table of EPROM addresses and their contents.

Input	Output	Input	Output	Input	Output
12	27	14	32	16	39
18	43	20	51	22	67
24	76	26	84	28	90

Exercise 4.5

Use EPROM, a binary counter, and gates to produce outputs labelled as water-in, water-out, detergent-in, wash, spin in the following sequence:

Water-in, wash, water-out, spin, detergent-in, water-in, wash, water-out, spin, water-in, wash, water-out, spin.

Would you expect a sequence like this to be useful if the pulse inputs to the counter were at equal time intervals?

Memory addressing

Practically all memory systems will include both ROM and RAM, so that memory must be mapped, meaning that it is connected so that one set of memory addresses will address ROM and another set of addresses will activate RAM. Some machines require the RAM also to be controlled so that a range of addresses put out by the microprocessor will activate a different set of addresses on the memory. In addition, though some ROMs use an 8-bit set of data pins, RAM memory chips mostly use 1 bit of data per chip; a few can use 4 bits. The mapping of memory is carried out using the chip-select pins on the ROM and RAM chips.

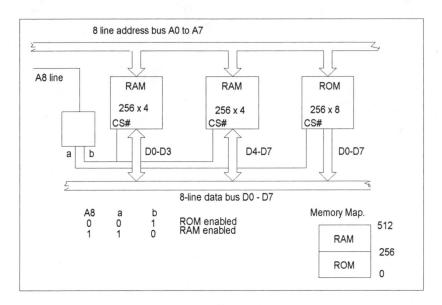

**Figure 4.14   Using RAM and ROM, mapped to different address ranges
by using the A8 line in this example**

Figure 4.14 shows a simple memory map and how it is achieved using a 256 x 8
ROM and two 256 x 4 RAM chips. The aim is that addresses 0 to 255 will
activate the ROM, and addresses from 256 to 511 will activate the RAM. Now a
256 set of data bytes will need eight address lines (because $2^8 = 256$), and these
will be address lines A0 to A7.

We do not want both sets of chips to be activated when signals exist on A0 –
A7, however, so the CS# (chip-select, activated by logic low) pins are used.
When the A8 line is low, the decoder will produce a low output at line a,
activating ROM, so that ROM uses the first 256 address numbers. When the A8
line is high, the decoder switches over and makes line b low and line a high,
deselecting the ROM and selecting the RAM chips. The next 256 address
numbers will therefore select RAM, and because the A8 line is high, these are
addresses 256 to 511, not 0 to 255. Of the two RAM chips, one contributes
D0 – D3, and the other D4 – D8, since the RAM chips are 4-bit data chips.

Note that if the microprocessor in such a system were allowed to use a full
address range, the alternate addressing of ROM and RAM would continue for
other memory ranges. For example, the (denary) range 512 to 767 would be
ROM and the range 768 to 1,023 would be RAM and so on. This repetition of
memory ranges can be avoided by using more elaborate decoding.

Take, for example, the arrangement of Figure 4.15, overleaf, which shows
ROM and RAM each of 4K (4,096 bytes), using address lines A0 to A11 (since

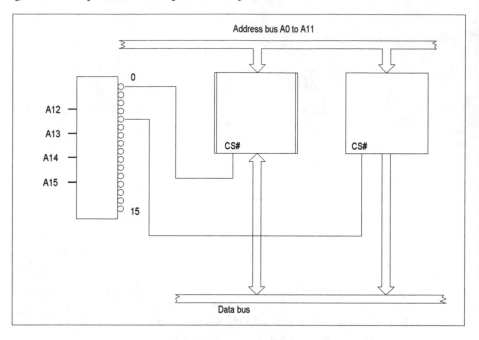

Figure 4.15   Address decoding for a pair of 4K chips, one ROM and the other RAM — in this example, the RAM chip is shown as 4,096 x 8, though it would more likely be a pair of 4,096 x 4 chips

2^{12} = 4096). The chip select pins in this example are driven by a decoder which uses the upper address lines as inputs, and gives outputs which go low for each possible state of the lines A12 to A15. The ROM is shown connected to the zero output of the decoder, corresponding to lines A12 to A15 all zero, so that the mapping of the ROM starts at address 0000H, and goes up to 0FFFH. The RAM is shown enabled by the number 4 output, corresponding to the line states A15 = 0, A14 = 1, A13 = 0, A12 = 0. This corresponds to the hex number 4000, so that RAM mapping in this system is at this starting address, and will extend to 4FFFH.

RAM on a computer (as distinct from controller) system is likely to use 1-bit chips, so that a 1 Mbyte memory will use 8 x 1M chips. Note that the byte is used to mean 8-bit memory even on machines which read and write in 32-bit Dword units, so that a 1 Mbyte machine with a 32-bit processor still means 1M x 8 of RAM, not 1M x 32. The connection of RAM for such purposes is straightforward, using each memory chip connected to its own data line for an 8-bit data bus, but when the data bus uses 16 bits, the memory must be split so that 16 chips

74

are used, with one chip to a bit as before. This system makes it easier to implement banking, with each 8-bit byte taken from a different set of chips.

Test Questions

1. Why are both ROM and RAM needed in a microprocessor system?
2. A computer is switched on and displays the time on its screen. How would you expect this to be done, assuming that no keys have been pressed?
3. If the program on a fusible-link EPROM has to altered, is there any way that this could be done?
4. What are the advantages of using a program or data on EPROM as compared to masked ROM?
5. Why does dynamic RAM need a refresh action? Why is static RAM not used to a greater extent for desktop computers?
6. Suppose that you wanted a sequence of input pulses to generate a set of 16-bit numbers, but you could use only 8-bit PROMs. How would you arrange your circuit to achieve this?
7. Figure 4.16 shows a system that uses ROM, EPROM and RAM. State the total storage capacity in bits, and write, in hex, the start and end address for each of the three sections of memory. Can you suggest a chip that might be used as the decoder in this system?

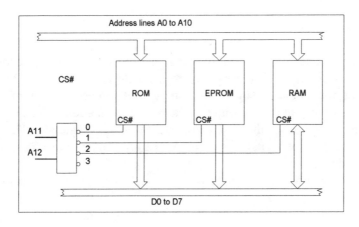

Figure 4.16   The system for question 7

8. What is the essential difference between EPROM and EAROM?
9. Describe how an EPROM can be programmed, naming a chip type.
10. Why is memory banking used for very fast computers with clock rates of 50MHz or more?

5   Analogue and digital conversion processes

Syllabus section: D04–4.1, D04–4.2, M04–4.1

Digital methods

Due to the rapid developments in digital signal processing and IC manufacturing technology, it is now much easier, more accurate and less costly to carry out many of the traditional analogue signal processes in the digital domain. This is particularly obvious in the measurement and instrumentation field, in spite of the fact that most signals in the real world are by nature analogue. The increased bandwidth needed for digital processing is readily compensated for by the many advantages that include:

1. A significantly higher transmission speed than can usually be achieved with analogue processing.
2. Improved transmission quality in noisy environments.
3. Better compatibility with the digital switching techniques used to control distribution making it a natural technique to use for systems using local area networks (LAN).
4. Encryption/decryption can easily be adopted for data security.
5. Where necessary, signal compression/bit rate reduction techniques can be employed to minimise the bandwidth requirement.

Sample and hold

The principle of converting the analogue signal into a digital format is explained in Chapter 6. This process is achieved by using a *sample and hold* IC, usually fabricated in MOSFET technology and using the principle shown in Figure 5.1.

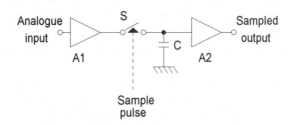

Figure 5.1   Basic sample and hold circuit

The analogue signal input is buffered by amplifier A1 to provide the low output impedance which is necessary to charge capacitor C quickly through the MOSFET sampling switch S. When S is closed, the circuit is in the sampling mode and the voltage across C follows or tracks the analogue input. When the switch is open, the circuit is in the hold mode and the buffer amplifier A2 couples the capacitor output voltage to the quantisation stage. This action is shown in Figure 5.2, from which it can be seen that between times t_1 and t_2, the acquisition time, the capacitor is charging at a rate determined by its value and the peak output current that can be provided by A1.

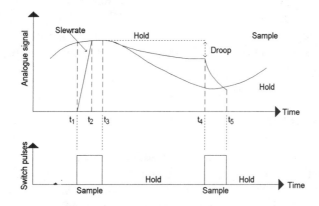

Figure 5.2   Sample and hold parameters

During this period, the voltage is changing at a slewing rate $dV/dt = I_{max}/C$, which can be as high as $3V/\mu s$. In a practical case, the charge on C leaks away somewhat during the hold period t_3 to t_4. This is due to the switch leakage, the dielectric absorption and the minute input current drawn by A2. This feature referred to as drift or droop, is also shown in Figure 5.2.

The drift or droop rate can be as small as 1mV/s. The value chosen for C, typically about 50nF, is a compromise. A high value reduces the droop, whilst a low value reduces the acquisition time and allows the sampled signal to track the input more accurately. The acquisition time quoted in converter specifications represents a worst case value and usually refers to the response to a maximum amplitude change.

An aperture time, typically in the order of 100ns, is also quoted in the specification for the sample and hold circuit. This represents the time delay between the onset of the switch pulse and the actual switch closure.

Important converter parameters

Resolution. This represents the smallest change that can be detected and is equal in amplitude to the least significant bit (LSB) = $V_{fs}/2^n$, where n is the number of bits.

Full scale voltage (V_{fs}). This is the nominal design maximum input amplitude. For most converters, the maximum output value will in general be lower than this depending upon the converter code and the number of bits. For example, for an 8–bit converter with $V_{fs} = 100V$, the resolution is $100/2^8 = 0.390625$ V and this would represent the 00000001 code value. The maximum input value of 100 V would be coded as 11111111.

Dynamic range. This is simply the ratio of the maximum voltage level to the resolution. In the above case this is $100/0.390625 = 256{:}1$, or $20\log 256 = 48.165$ dB. Since a doubling of voltage represents a change of about 6dB, the dynamic range is also approximately equal to 6n dB, where n is the number of bits.

Quantising ac signals

The quantising process described above, is satisfactory for unipolar signals with a large dc component. For ac or bipolar signals, an alternative approach is needed. One method involves adding a constant to each sampled value, using the offset-binary technique. But in those cases where signals from different sources have to be added together, the sum can overflow and exceed the allowable peak values. A common solution involves the use of the 2's complement method of representing a binary number. By convention, a leading 0 then indicates that the remaining code represents a positive number, while a leading 1 signifies a negative quantity. The following gives partial examples of these three coding methods.

Straight binary coding. The sequence of digits are simply binary weighted in the following manner. The analogue voltage (V_{an}) is given by: $V_{an} = V_{fs}(\frac{1}{2} + \frac{1}{4} + \frac{1}{8} + \dots + \frac{1}{2^n})$. For a 4-bit converter with $V_{fs} = 10V$, the following listing gives some of the code values.

$0\ 0\ 0\ 0$	$= 0V$
$0\ 0\ 0\ 1 = 10\ (0 + 0 + 0 + \frac{1}{16})$	$= 0.625V$
$0\ 0\ 1\ 0 = 10\ (0 + 0 + \frac{1}{8} + 0)$	$= 1.25V$
$0\ 0\ 1\ 1 = 10\ (0 + 0 + \frac{1}{8} + \frac{1}{16})$	$= 1.875V$
...	...
$1\ 1\ 1\ 0 = 10\ (0 + \frac{1}{2} + \frac{1}{4} + \frac{1}{8})$	$= 8.75V$
$1\ 1\ 1\ 1 = 10\ (\frac{1}{2} + \frac{1}{4} + \frac{1}{8} + \frac{1}{16})$	$= 9.375V$

Binary offset coding. To cater for ac signal voltages, the binary pattern is level shifted so that the value for half full scale is set to zero. The analogue voltage is given by; $V_{an} = V_{fs}(1 + \frac{1}{2} + \frac{1}{4} + \dots \frac{1}{2^{n+1}} - 1)$. For a 4-bit converter with $V_{fs} = \pm10V$, the same code values now yield:

$0\ 0\ 0\ 0 = 10\ (0 + 0 + 0 + 0 - 1)$	$= -10V$
$0\ 0\ 0\ 1 = 10\ (0 + 0 + 0 + \frac{1}{8} - 1)$	$= -8.75V$
$0\ 0\ 1\ 0 = 10\ (0 + 0 + \frac{1}{4} + 0 - 1)$	$= -7.5V$
$0\ 0\ 1\ 1 = 10\ (0 + 0 + \frac{1}{4} + \frac{1}{8} - 1)$	$= -6.25V$
...	...
$1\ 1\ 1\ 0 = 10\ (1 + \frac{1}{2} + \frac{1}{4} + 0 - 1)$	$= +7.5V$
$1\ 1\ 1\ 1 = 10\ (1 + \frac{1}{2} + \frac{1}{4} + \frac{1}{8} - 1)$	$= +8.75V$

In this case, the resolution is given by $20/2^4 = 1.25$ volts.

2s complement coding. In this case the analogue voltage is given by: $V_{an} = V_{fs}(-1 + \frac{1}{2} + \frac{1}{4} + \frac{1}{8} \dots + \frac{1}{2^{n-1}})$. For the same 4-bit pattern and $V_{fs} = 10V$, the same code values now yield:

$0\ 0\ 0\ 0 = 10\ (0 + 0 + 0 + 0)$	$= 0V$
$0\ 0\ 0\ 1 = 10\ (0 + 0 + 0 + \frac{1}{8})$	$= +1.25V$
$0\ 0\ 1\ 0 = 10\ (0 + 0 + \frac{1}{4} + 0)$	$= +2.5V$
$0\ 0\ 1\ 1 = 10\ (0 + 0 + \frac{1}{4} + \frac{1}{8})$	$= +3.75V$
...	$= ...$
$1\ 1\ 1\ 0 = 10\ (-1 + \frac{1}{2} + \frac{1}{4} + 0)$	$= -2.5V$
$1\ 1\ 1\ 1 = 10\ (-1 + \frac{1}{2} + \frac{1}{4} + \frac{1}{8})$	$= -1.25V$

Exercise 5. 1

Complete the tabulated values for the straight binary, binary off-set and 2s complement coding methods quoted above.

Analogue to digital (A/D) conversion processes

Counting or voltage-to-time A/D converter

The clock circuit of Figure 5.3 provides a source of precisely timed pulses. After the count has been set to zero by the Clear or Start signal, the pulse count is accumulated in the binary counter.

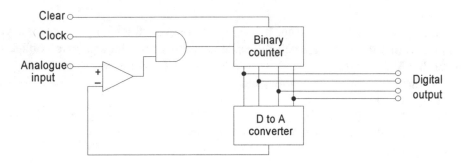

Figure 5.3   Counting or Voltage-to-time A/D converter

This count provides the digital output and is also used to provide a comparison signal after being converted into analogue form. As long as this value remains below the level of the analogue input, the AND gate is enabled so that the count can accumulate. As soon as the comparator inverting input exceeds that of the analogue input, its output goes negative, the AND gate is disabled and the count ceases. The value held in the binary counter can then be output as the digital representation of the input analogue signal. As with most converters, due to the repetitive nature of the counting process, the last digit will fluctuate by one count.

Single slope or voltage to frequency A/D converter

In the system shown in Figure 5.4, clock pulses pass through the gate circuit to trigger a pulse generator. The pulse widths are defined by these pulses and their amplitude controlled by the reference voltage. Thus when filtered, these pulses provide a dc input to the comparator that is proportional to the amplitude of the reference.

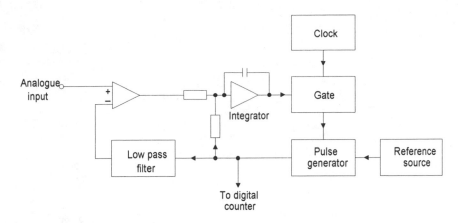

Figure 5.4   Single-slope or voltage-to-frequency A/D converter

The analogue input is compared with this level to provide a drive for the integrator and as this output rises the gate opens to allow the pulse generator to be triggered by the clock. At the appropriate times, the pulse generator also resets the integrator. The feedback system balances the number of generated pulses against the level of input signal. When these pulses are digitally counted, they represent the digital conversion of the input signal. The circuit is also self compensating for clock frequency drift. If the clock frequency rises, the comparator input will tend to fall so that the count is performed in a shorter period of time. The number of pulses is thus relatively unchanged.

The acquisition time is signal amplitude and clock frequency dependent. However, since the D/A converter changes by the resolution at each clock pulse, the slew rate is given by the ratio of the resolution to the clock periodic time. For example, for an 8-bit converter with $V_{fs} = 10V$, the resolution is $10/2^8 = 39mV$. For a clock rate of 1MHz (period = 1µs), the slew rate is 39mV/s. For $V_{an} = 1V$, the acquisition time is $1/(39 \times 10^{-3}) = 25.65µs$, whilst for a 20V input this becomes 513µs. Thus if the clock frequency is doubled the acquisition time is halved. If the maximum amplitude voltage can be represented by n pulses when the clock period is T seconds, then the conversion time is given by nT seconds.

Dual-slope or ratiometric A/D converter

With the method depicted in Figure 5.5(a), the sequence starts with the analogue input signal applied via the switch to an integrator. As soon as the integrator output starts to rise, the gate opens and clock pulses are passed to the counter, which has been preset to some value and now starts to count down. When the count reaches zero, a signal is passed to control and the integrator input now switches to a reference voltage of the opposite polarity. This causes the capacitor

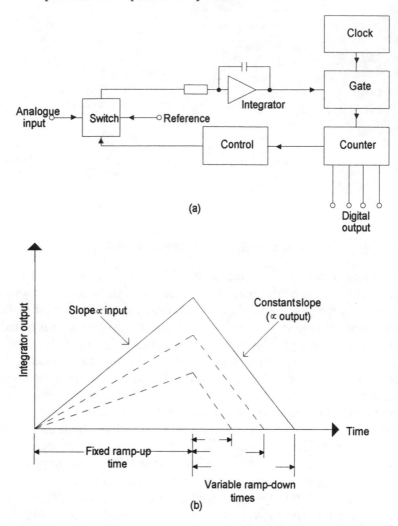

(a)

(b)

Figure 5.5   Dual-slope or ratiometric A/D converter: (a) block diagram, (b) action

to discharge but at a constant current, with the counter counting upwards during this period. The time taken to discharge the capacitor is proportional to the level of input voltage originally applied. Thus the count at this point provides a digital equivalent of the input signal.

From Figure 5.5(b), it will be seen that the ramp-up period is fixed so that the accumulated charge is proportional to the amplitude of input signal. As the discharge period occurs at a constant current, the ramp-down time must be proportional to the charge on the capacitor and hence the level of the input signal.

Clock frequency drift is relatively unimportant in this system, as the voltage levels are dependent more upon the number of clock pulses than their rate. All other parameters being equal, the improved accuracy is achieved for a slightly lower conversion rate than the single slope converter.

Flash converter

This aptly named, very fast, technique is shown in Figure 5.6(a). The analogue input is applied simultaneously to a number of parallel comparators. For M levels this system requires M–1 comparators and b bits to encode the digital output $(M = 2^b)$.

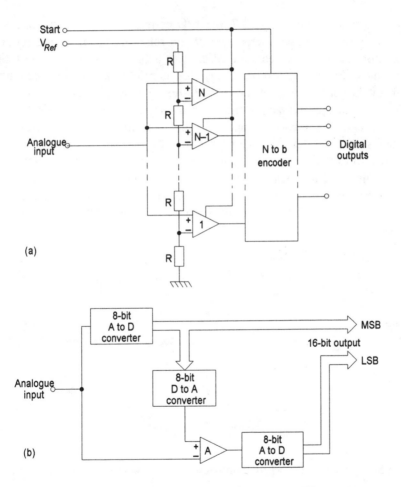

Figure 5.6   **Flash converters: (a) full converter, (b) half-flash converter**

Digital Techniques and Microprocessor Systems

The second input to each comparator is obtained via a potential divider network so that appropriate multiples of $1/M^{th}$ of the reference voltage is obtained. The parallel outputs from the comparators are then encoded to provide the digital equivalent of the input signal.

The half flash conversion system shown in Figure 5.6(b) provides a convenient way to double the number of output bits. The input is first quantised to provide the eight most significant bits (MSB) and this is then converted back into analogue form to be subtracted from the original signal. When this difference in converted into digital form, it represents the eight least significant bits (LSB).

Successive approximation A/D converter

In the system shown in Figure 5.7, the A/D converter acts a as digital divider, to present accurately known increments of the reference level to one input of the comparator. These increments are offered in a fixed sequence, largest first and decreasing in binary proportion. If any step is offered and it produces a running sum that is larger than the analogue input, this value is rejected and the next tried. The successive approximation register thus accumulates a total in binary form that represents the analogue input. As the sequence ends, the counter stage commands the register to output its contents. A start pulse is used to start conversion and a delay circuit is used to provide automatic recounting. These pulses being used to open the clock gate to provide the necessary drive to the counter circuits. Often the successive approximation register outputs provide the digital signal in serial format, with a status flag to signify that counting is in process.

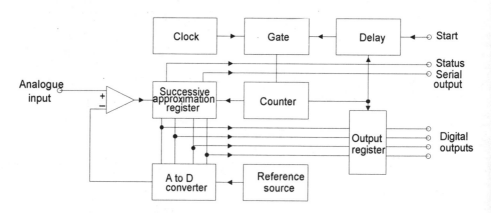

Figure 5.7   Successive approximation A/D converter

For this type of converter, the acquisition time is relatively independent of the input amplitude and is approximately given by the product of the number of bits and the clock period. Typically a 12-bit converter has an acquisition time as low as 1μs.

Exercise 5.2

Operate A/D converters at a low clock frequency; measure the differences in conversion times between the counter ramp and the successive approximation methods. Observe on an oscilloscope the aliasing effects of applying an input sinewave with a frequency equal to that of the converter bandwidth. Observe the digital outputs for various analogue input levels and evaluate the quantisation error.

Digital to analogue (D/A) conversion processes

Binary weighted resistor network

A series of parallel transmission gates provide inputs to an inverting operational amplifier (op-amp) that functions as a summing amplifier. Reference to Figure 5.8 will show that these gates switch the inputs to either earth potential or to a stable negative reference voltage, $-V_{Ref}$. A binary zero on any data line connects that input to 0 volts, whilst a 1 sets the input to $-V$ volts.

The resistor network is chosen in binary proportions as shown, so that the gain of the op-amp can vary as the series $1/2$, $1/4$, $1/8$, $1/16$, etc. The negative reference is chosen so that the analogue output V_{out} will be positive. Thus the analogue output signal will be: $V_{out} = V_{Ref}(A_1/2 + A_2/4 + A_3/8 + ...)$ where A_1, A_2, etc are set to 1 or 0 according to the bits in the input binary word. A low pass filter is required to remove the quantisation steps that will be present at the op-amp output. The chief disadvantages of this system lie in the resistor values which must be very accurate, stable and track each other as the operating temperature changes. Also if there are a large number of bits in each binary word, some resistor values become very large and this adds to the stability problem as well as tending to slow the rate of conversion. Because of these problems, binary weighted converters are used for non-critical applications.

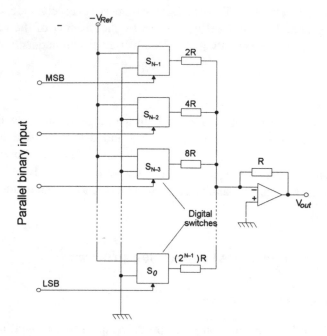

Figure 5.8   Binary weighted resistor network D/A converter

R/2R ladder network D/A converter

Each leg of the ladder network shown in Figure 5.9, can be connected either to ground or the reference voltage, by the transmission gates which are controlled by each bit of the input binary word. The network is a current splitting device that due to the R/2R ratio of the resistors produces a sequence of binary ratios.

Analysis of the circuit shows that a resistance of 2R is seen from any node, looking left, right or towards the switches.

> At node N the op-amp gain = $-3/2$,
> Voltage due to V_{Ref} at N = $-V_{ref}/3$,
> So that $V_{Out} = -V_{Ref}/3 \times -3/2 = V_{ref}/2$
> Similarly at node N–1, $V_{Out} = V_{Ref}/4$, etc.

This again produces the ratios $1/2$, $1/4$, $1/8$, etc, with the op-amp acting as a summing amplifier. Temperature stability with this network is easily achieved, as only two resistor values in the 2:1 ratio are needed for accuracy.

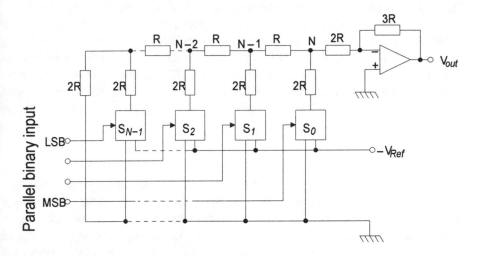

Figure 5.9   R/2R ladder network D/A converter

Current driven D/A converters

Very high speed conversion systems need to operate with low impedance circuits and this can be best achieved by using binary weighted currents. A good example of this concept is the Plassche converter which is based on the principle of current addition shown in Figure 5.10(a) where $I_3 = I_1 + I_2$. For accuracy, the resistors and current values should have a close tolerance. However, in the Plassche converter this problem is eased by introducing a high-speed switching action as shown in Figure 5.10(b).

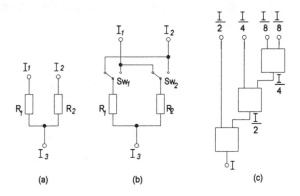

Figure 5.10   The Plassche converter: (a) current divider, (b) switched divider, (c) cascaded stages

87

Digital Techniques and Microprocessor Systems

Here the currents I_1 and I_2 are alternately switched between the two resistors so that over a short period of time and by adding a smoothing capacitor, any differences average out to zero. Variations of resistor tolerances in the order of 1% can readily be accepted. It is however important that the clock signal should have an accurate 1:1 mark to space ratio. Figure 5.10(c) shows how a minimum number of these modules can be integrated to make an efficient high-speed converter.

Multiplying D/A converters

From the above it will be seen that the D/A converter output is the product of the reference voltage and the binary word and always less than the reference voltage. If the fixed reference is replaced with an analogue signal, the device becomes a *Multiplying* D/A converter or *Programmable Attenuator*, with the degree of attenuation being controlled by the digital word. The concept of using one dc and one ac voltage is referred to as a *2-Quadrant Multiplier*. The combination of a bipolar binary signal with an ac reference forms a *4-Quadrant Multiplier*.

Converter errors

Because both A/D and D/A converter circuits contain amplifiers and comparators, there is scope for distortion, noise and temperature drift to introduce errors. Of these the most troublesome are as follows:

Linearity. As shown in Figure 5.11(a), the ideal characteristic is represented by a straight line and any deviation from this due to non-linearity, results in non-equal outputs steps. The magnitude of error is usually expressed as a percentage of the full scale value or fraction of the LSB. Since quantisation errors can account for (LSB), a good converter should not exceed this.

Offset Error. An uncalibrated converter could produce a response as shown in Figure 5.11(b) where the curve does not pass through the zero point. The circuit

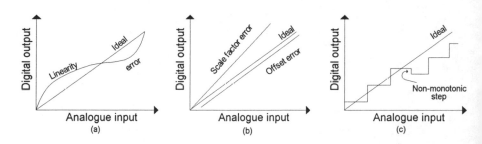

Figure 5.11 Converter errors: (a) linearity, (b) scale factor and offset, (c) non-monotonicity

calibration involves making an adjustment to a preset control to correct for this offset.

Scale Factor or Gain Error. Again Figure 5.11(b) shows how the response curve may have a different slope to the ideal. This is due to gain variation which can be corrected during calibration.

Monotonicity. A monotonic series of numbers is represented by a continually increasing or decreasing set of values. In a monotonic converter, the series of steps are either increasing or decreasing continually between certain limits. If a glitch or interference pulse generates a step of the wrong polarity as shown in Figure 5.11(c), then the current count will be in error. Furthermore, non-monotonicity can arise from an error in the bit weighting network to give rise to missing codes. The count can occur correctly up to the point at which the retrograde step occurs, but the code for the next step above this will never be output.

Test Questions

1. (a) State the feature that determines the resolution of an A/D converter.
    (b) Calculate the highest component frequency present in a signal that when sampled at 8 bits per conversion cycle, can be handled by a converter with a 2μs acquisition time, without exceeding the bandwidth.
    (c) For an 8-bit counter/staircase A/D converter, calculate the maximum and minimum conversion times if the clock is running at 10kHz.
2. An A/D converter is used to convert an analogue signal with a frequency range of 3kHz to 8kHz.
    (a) State the minimum theoretical sampling frequency for alias free conversion.
    (b) State the typical practical sampling frequency.
    (c) Calculate the minimum number of bits for a resolution better than 0.1%.
    (d) Calculate the error when an A/D converter with 100ns aperture time is driven with an input signal varying linearly at 2V/s.
3. An 8-bit A/D converter samples an analogue signal once every millisecond. It produces the codes 0000,0000 and 1000,0000 with zero and +5V inputs respectively. Calculate:
    (a) The dynamic range at the input.
    (b) The number of quantisation levels.
    (c) The voltage represented by each level.
    (d) The output bit rate.
    (e) The maximum theoretical alias free conversion rate.

4. An A/D converter is used to convert a 10kHz analogue signal with a resolution of 0.1% using the minimum theoretical alias free sampling frequency. Calculate:
    (a) The minimum number of bits for each sample.
    (b) The number of quantisation levels needed.
    (c) The sample clock frequency.
    (d) The output bit rate.
    (e) The minimum theoretical bandwidth of the converted signal.

5. (a) Sketch the circuit of a 4-bit weighted resistor D/A converter that provides an inverting output.
    (b) If the input resistors are 1k, 2k, 4k and 8k, indicate which provides the LSB and the MSB inputs.
    (c) Calculate the value of the feedback resistor to give an output change of –1v when +4V is applied to the LSB input with all the others set to zero.

6   (a) The circuit sketched in Q5(a), has to produce an output of twice the MSB input when all the other inputs are set to zero. If the feedback resistor is 1k, calculate the values of the four input resistors.
    (b) A 4-bit R/2R ladder network D/A converter produces outputs of 0V and +4V for the input codes, 0000 and 1000 respectively. Calculate the number of quantisation levels and the resolution.
    (c) State two advantages of the R/2R device over the weighted resistor network converter.

6 Signalling and modulation

Syllabus references: B1–E.6, B2–C.6, M02–2.1, M09–9.1.

Digital communication systems

Wideband communications applications have traditionally used analogue signal processing, primarily for reasons of bandwidth conservation and the fact that the technology is matured and well understood. However each application tends to be unique in certain ways. When such systems are concentrated into integrated circuits (ICs) these devices become specialised, relatively few are made, and so their costs are higher than they would be if they were mass-produced. When the analogue signals are converted into digital form for processing and then back into analogue again for output, the only dedicated ICs are those associated with the interfaces between the two types of signal. The digital signal processing (DSP) region of the system then uses standard digital components that are mass-produced so that the system becomes more cost effective.

   Digital signals are often characterised by their structure and information transfer rate. For convenience binary digits (bits) are organised into a hierarchy of various sized groups. 4 bits form a *nibble* that can represent one hexadecimal character. 8 bits form a *byte* or *octet* which can be used to represent an alpha-numeric or control character from the 7-bit based ASCII code set (the extra bit is commonly used for parity check). A *word* may consist of several bytes, often 2,

91

4, 8 or more long. The data transfer rate may then be quoted in bits, bytes or characters per second. Alternatively the term Baud rate may be encountered and this can cause confusion. By using suitable modulation techniques, it is possible that each transmitted symbol can represent more than 1 bit of information. If each symbol represents 4 bits, then the Baud or symbol rate is ¼ of the bit rate. Only if each symbol represents 1 bit are the bit and Baud rates equal.

The increased transmission bandwidth required for digital signals may be available over certain types of link, when this penalty can be offset by the considerable advantages of digital processing. The system becomes more flexible, different systems can be integrated and computer control introduced. Such a technique leads to the concept of an integrated services digital network (ISDN) where each service can be accommodated with equal performance. The principal benefits of digital processing can be summarised as follows:

1. It is more appropriate for linking devices that operate in the digital mode.
2. It can provide a significantly higher transmission speed than can usually be achieved with analogue processing.
3. It provides for improved transmission quality in noisy environments. The noise component can be reduced using signal regenerators and error detection/correction techniques.
4. It is more compatible with the digital switching techniques used to control distribution and is a natural technique to use for systems involving an optical fibre link.
5. Encryption/decryption can easily be added for data security.
6. Where necessary, signal compression/bit rate reduction techniques can be employed to minimise the bandwidth requirement.
7. For many applications, time division multiple access (TDMA) can be used more effectively than frequency division multiple access (FDMA) that is common for analogue transmission systems.
8. For systems involving reception and retransmission, signal regeneration can be used at the intermediate stage to improve signal quality.

Communicating terminals may operate in one of two modes. In the *half duplex* mode, the terminal can only transmit or receive alternately. By comparison a *full duplex* terminal can transmit and receive simultaneously. Terminal devices that operate in the text mode such as printers, keyboards, visual display units (VDUs), etc., often communicate through the ASCII code table. For videotext/teletext operations a variant of this code set is used which provides for graphics characters. This is known as the ISO-7 code (International Standards Organisation — 7-bit code). Since 7-bit codes provide only for 128 different character symbols, an extended code set, EBCDIC (extended binary coded decimal interchange code) providing for 256 different alpha-numeric, graphics and control symbols is available for special purposes.

Pulse code modulation (PCM) system

The basic principle of this form of digital communications is shown in Figure 6.1(a) which indicates how an analogue signal is converted into a digital format. This process is described as sampling and quantisation. The analogue signal is sampled (measured) at very precise intervals of time (sampling intervals), to evaluate its amplitudes. As only discrete values are used in a digital system, each of the sampled values is quantised or rounded to the nearest lower integer value.

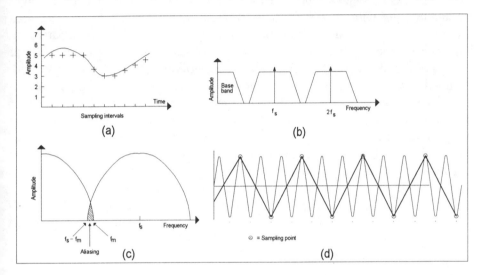

Figure 6.1 **PCM system coding: (a) sampling and quantisation, (b) frequency spectrum due to sampling, (c) aliasing (frequency domain), (d) aliasing due to slow sampling (time domain)**

The waveform shown in Figure 6.1(a) would thus be represented by the binary sequence: 100,101,101,101,101,100,011,010,010,011,100... It will be seen that each of the eight levels can be coded with just 3 bits. The general rule for binary coding is $M = 2^n$, where M is the number of discrete levels and n is the number of bits per sample.

Nyquist's sampling theorem shows that if a complex analogue signal is sampled at a rate of at least twice that of its highest frequency component, the original signal can be recovered from these samples without error. This arises because the sample points tend to regenerate a square wave, which when low pass filtered yields the original signal. However, errors will occur because of the approximation produced by quantisation. This is described as quantisation noise which in practice, varies between zero and $\pm\frac{1}{2}$ (LSB).

Reference to Figure 6.1(a) shows that quantisation noise can be reduced to a low level simply by increasing the sampling rate and or the number of allowable levels. The cost of this being an increase in the bandwidth of the digital signal. This can be calculated from $2nf_m$, where f_m is the maximum component frequency and n is the number of bits per sample. Thus for a baseband analogue signal extending to 10kHz, the sampling frequency should be at least 20kHz. If 8 bits per sample are needed, then the bit rate becomes 8 x 20kHz = 160 kbit/s, so that the minimum transmission bandwidth would be 80kHz.

The process of sampling produces a frequency spectrum similar to amplitude modulation, but with an infinite range of harmonics as shown in Figure 6.1(b). The receiver demodulator circuit must include a low pass filter to separate the baseband component from the harmonics. If the sampling frequency is not high enough or the filter cut-off not sharp enough, interference from the first lower side band will result. This effect known as *aliasing* due to use of a sampling frequency below the Nyquist value is shown in Figure 6.1(c). In practice, it is often necessary to low pass filter the analogue signal before sampling to ensure that noise or distortion components will not generate high frequencies that would create aliasing. Such a device is then referred to as a Nyquist filter. Furthermore, any very small analogue signals may need amplification before digital processing. In which case, it is important to filter off any extraneous noise that would tend to mask these small signals. The fact that aliasing produces signal components that were not present in the original signal can be clearly seen in the time domain diagram of Figure 6.1(d).

The integrated services digital network (ISDN)

An ISDN is a communications network that has evolved because of the problems associated with analogue telephony systems. The concept recognises the considerable advantages that can be gained by changing to a system that will allow the end-to-end transfer of information in a digital manner. Once such a transition has been made, the advantages gained include:
1.   Greater reliability due to the use of digital integrated circuits.
2.   Faster access speed due to the introduction of dual tone multi frequency (DTMF) dialling instead of pulse dialling (see Table 6.1).
3.   Allows for computer access to the system which gives rise to the concept of computer integrated telephony (CIT), where the power of the computer can be combined with the communications power of ISDN.
4.   By introducing CIT, many new services such as video, data and Group 4 high-speed facsimile can be introduced into the telephony system in a truly integrated fashion.

Table 6.1   DTMF dialling tones

	1209	1336	1477
697	1	2	3
770	4	5	6
852	7	8	9
941	*	0	#

All frequencies in Hertz. Push buttons and cross point switches
select two frequencies, from one above, and one below 1kHz
according to the above matrix

For reasons of economy, the service has to be compatible with current analogue systems and introduced in an evolutionary manner. Thus voice signals are processed in analogue to digital and digital to analogue converters to be carried over the new system by PCM.

ISDN offers two classes of service, a basic rate which most subscribers use, known as 2B+D (bearer + diagnostic), consisting of two 64Kbit/s voice and data channels, plus one 16Kbit/s digital signalling channel making a total of 144Kbit/s. The former is circuit switched, whilst the latter is packet switched. The second, more costly service consists of 30B+D channel groupings in Europe (CCITT standard, 32-channel multiplex) and 23B+D (Bell standard, 24-channel multiplex) in North America, giving total signalling rates of 2.048Mbit/s and 1.544Mbit/s respectively. The D channel forms a *common signalling channel* for the group of bearer channels. Control is exerted through the D channel in the following manner. The calling party transmits the called number plus other information, then, if just a voice connection between two digital telephones is needed, only one 64Kbit/s B channel will be required. However if voice plus data is involved then two B channels must be engaged from end to end. The calling party thus negotiates the necessary channel capacity using time sharing of the D channel. When not needed for call routing control, the D channel can be used for packet switched data or network monitoring.

Access to the packet switched network is obtained via a *packet assembler/disassembler* (PAD) which is placed between the ISDN and the packet switched public data network (PSDTN). For the packet switched system, the data is organised into blocks or packets in the manner shown in Figure 6.2, where:

(a)   Start flag indicates the beginning of the packet and can contain a preamble sequence of bits designed to aid synchronisation.

(b   Each packet must contain the address of the originator and the destination.

(c)   Control section holds the packet sequence number that is needed to reassemble the message in the correct order if it exceeds one packet in length.

(d)   The data bits.
(e)   A frame check can be used to provide error protection and correction.
(f)   An end flag to signify the last bit in the packet.

Start	Address	Control	User Data	Frame check	End
(a)	(b)	(c)	(d)	(e)	(f)

Figure 6.2   Data packet structure

When the user has obtained access to the network using the appropriate identification, he specifies a destination address and then inputs the message. The PAD breaks this up into suitable blocks to form the packets and automatically adds the controlling information before transmission to the network. Should the receiving end detect an error, then a *repeat transmission* request will be generated.

Digital modulation schemes

In order to increase the number of communication channels and maximise the throughput of data, digital baseband signals are often modulated onto high frequency carrier waves. Since any sinewave can be modified or distorted by a baseband signal in three ways, the amplitude, frequency or phase information of the carrier can be used to convey the data signals. These characteristics give rise to the following methods of digital modulation.

With *amplitude shift keying* (ASK) the amplitude of the carrier is modulated according to the discrete level of the data signal. For binary modulation, the carrier is simply switched ON or OFF to represent 1 or 0 respectively. This technique is also known as on-off-keying (OOK). Because ASK and noise signals have similar characteristics, this form of modulation is not popular because of its poor S/N ratio and hence relatively high bit error rate (BER).

Binary *frequency shift keying* (FSK) involves switching the carrier wave between two frequencies. One form of this is known as Kansas City modulation for which bursts of eight cycles of 2400Hz or four cycles of 1200Hz are used to represent 1 or 0 respectively. Other frequency values are used and these may or may not have an exact 2:1 ratio. Small differences are often introduced so that beat notes between the two frequencies will not introduce bit errors. FSK has a better BER performance than ASK under the same conditions, but requires a wider transmission bandwidth.

Phase shift keying (PSK) is a single frequency method in which the data signal is used to switch the carrier phase. Typically for binary transmission a 0 produces no effect whilst a 1 generates 180° of carrier phase shift. Of the three methods, PSK has the best BER performance and the narrowest transmission bandwidth.

Multi-phase PSK, in which the modulated carrier can carry a number of different phase shifts, can be used to advantage. Figure 6.3(a) shows an example of 8-PSK where each vector can be used to represent 3 bits of information so that the Baud rate is just $^1/_3$ of the bit rate. The concept of 4-PSK or *quadrature PSK* (QPSK) can be usefully extended in several ways. For example, if each of the four vectors are permitted to have any one of four different amplitudes, then each vector can be used to represent 4 bits. Figure 6.3(b) shows one example of such *quadrature amplitude modulation* (16-QAM) in which the signal points in the matrix form a constellation.

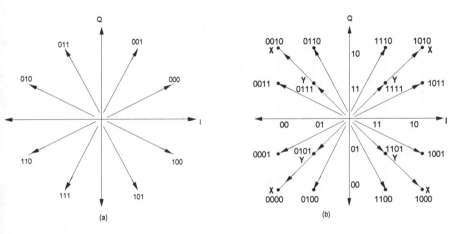

Figure 6.3   Digital modulation: (a) 8-PSK, (b) 16-QAM

Exercise 1

Use a modem to establish communications between two computers over a simple telephone extension network. Interpose a break-out-box between one computer and the modem to study the data signals at different communication rates.

Test Questions

1.   Briefly describe the important ways in which the ISDN system differs from the public switched telephone network (PSTN).

2.   Describe any advantages of the dual-tone multi-frequency (DTMF) keypad system over the older pulsed dialling device.

3.   Why is ASK signalling not a preferred method. What are the significant advantages of PSK signalling. State a common application of the use of FSK.

4.   Describe the CUTS (computer users transmission standard) or Kansas City modulation system. What are the significant advantages of this method.

7   Display systems

Syllabus section: D0–7.1 to D07–7.5, I1–3.7, I2–2.1, M0–3.1, M04–4.2.

The operation and control of the visual display units (VDU) or monitors associated with computer systems has much in common with the television systems and receivers that are extensively described in Volume 2, Part 2, of this series. To avoid a great deal of repetition, the reader is therefore referred to that book for any necessary revision.

The cathode ray tube (CRT)

The operating principles of this device, which forms the heart of most VDUs, are exactly the same as that of the television display tube. The CRTs designed specifically for graphics displays are built with a closer spaced, finer phosphor dot structure in order to improve the resolution. For slotted mask tubes the slots and phosphor stripes are made narrower. A shadow mask dissipates a significant degree of heat which can give rise to distortion and generate beam landing errors that in turn give rise to false colouration. Therefore the shadow mask is usually made from Invar, a nickel/iron alloy that has a very low coefficient of thermal expansion. Alternatively the tube may then be driven harder to produce a brighter display. The use of a face plate with a lower light transmittance, provides a blacker background to improve the contrast range and at the same time, reduce

Digital Techniques and Microprocessor Systems

X-ray radiation. The effects of beam landing errors can be further minimised with slotted shadow mask tubes by laying down black stripes between the groups of R, G, and B phosphors. The finer image detail necessarily requires a wider bandwidth for the video amplifiers. Incidentally, the shadow mask, delta gun tube often provides the highest degree of resolution. However, unlike the television receiver, the raster scanning is commonly sequential (non-interlaced) and employs somewhat different frequencies.

The typical operating voltages used for both monochrome and colour display tubes are basically the same as those used for television. Therefore exactly the same warnings of dangerous voltages associated with monitor CRTs still apply. In particular, never attempt to measure the EHT voltage with an ordinary meter.

Exactly the same geometrical image errors occur and these need to be corrected in exactly the same way following manufacturers instructions.

Display of colour images

It will be recalled from earlier work, that the additive colour mixing of three primary colours red (R), green (G) and blue (B) is used for generating computer and television type of colour images. Furthermore, most of the colours that occur in nature can be created by adding these primary colours together in suitable intensities. In this manner, three secondary colours — magenta (M), cyan (C), yellow (Y), plus white (W) — can be produced in the following way:

$$W = R + G + B$$
$$M = R + B$$
$$C = B + G$$
$$Y = G + R$$

Figure 7.1 shows how this colour mixing is achieved in a typical precision in-line (PIL) colour CRT. The video drive signals are dc coupled from the separate amplifier stages to the R, G and B cathodes. The typical average cathode voltages are about 90 to 100 volts, with the grid at near zero volts.

The grid electrode in this case provides for both user brightness control and beam blanking. The first anode voltage (A1) is presettable and typically operates at around +350 volts. The focus voltage which is also presettable, typically over the range of 4 to 6kV. The final anode voltage (EHT) which is typically set by specific adjustment to 25kV, is also connected to the shadow mask and the internal graphite coating within the tube bulb. The external graphite coating on the tube bulb must always be properly earthed, as this together with the glass and the internal coating, provides the smoothing capacitor for the EHT voltage.

Spark gaps are provided at strategic points in the circuit to protect external transistors from the transient effects due to flash over within the CRT. Such occurrences which may rupture the semiconductor junctions, can arise from

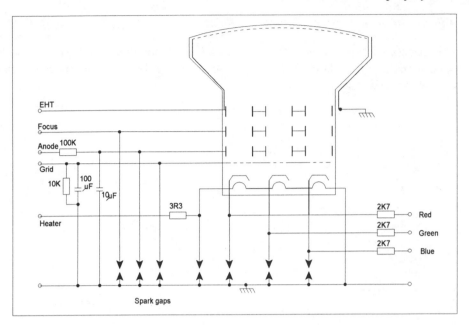

Figure 7.1   Circuit of typical colour CRT and drive inputs

sudden changes in the level of the EHT voltage or from rapid changes in the brightness level due to image information.

Faults that develop in the circuit of Figure 7.1 can produce some very interesting images. For example, if the Red signal fails due to loss of drive or an open circuit feed resistor, the image colours will change. White areas turn cyan, yellow areas turn green, magenta areas turn blue and red areas turn black. Similarly if the blue gun signal fails, white areas turn yellow and blue areas become black. An internal short circuit between a grid and cathode, leads to uncontrollable brightness and if this occurs on the red gun, the displayed image will have strong green over cast. Similar effects occur with the other two guns.

CRT adjustments

Note that before any adjustments can be effectively made, the monitor should be switched for a long enough period to allow the circuits to reach a normal operating temperature. Also always refer to the manufacturer's instructions first.

Adjustment of contrast, brilliance and focus

The effects of these three controls tend to be interactive and so need careful adjustment using a standard input test waveform referred to as a Grey scale, similar to that shown in Figure 7.2.

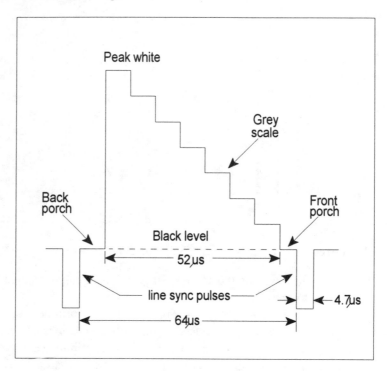

Figure 7.2   Grey scale luminance waveform (typical European timings)

These adjustments which apply equally to both monochrome and colour tubes should be carried out as follows:

1.   For a colour tube, turn down the colour control to leave a monochrome image.
2.   Set the contrast control to minimum.
3.   Adjust the brightness control so that black areas are just acceptable black. Then set the contrast control so that there is a good contrast ratio between the black and white stripes, with all the grey stripes being clearly distinguishable.
4.   Adjust the focus control so that the vertical edges of the grey scale pattern are sharply defined. The degree of resolution with this adjustment is governed to some extent by the bandwidth of the video drive amplifiers, hence the focus control may be found to have limited effect.

A correctly adjusted image should display black areas that are acceptably black without any visible scan lines, display clear highlights and with a good range of grey tones in between. A too-high setting of the brilliance control produces an image lacking in true black, whilst a low setting results in the loss of grey areas.

5.   Reset colour control.

Computer generated test waveforms

Whilst it is common to use a standard video waveform generator to provide the necessary alignment signals, it is possible to use the parent computer to generate these in software. Appendix 1 lists just three examples, written in BASIC for the PC machine, that are available on disk (courtesy of Ken Taylor).

Program 1 provides a screenful of Xs which enables the focusing to be checked over the whole screen. Program 2 provides red, green, blue or white rasters which are useful for purity adjustments and general signal tracing. Program 3 provides a circle and grid pattern for height, width, image linearity and centring adjustments. The program provides for either CGA or EGA modes.

Operation of visual display unit (VDU)

Just as the colour TV camera generates analogue video signals for the TV receiver, similar signals can be generated for the digital computer system display. Colours occurring in nature can be synthesised by combining brightness (luminance) and colour (chrominance) signals. Colour itself has two components, hue the actual colour and saturation, the degree by which intense colours are diluted with white light. Desaturated colours are often referred to as pastel shades. In order to obtain colour displays, it is necessary to use the CRT to convert these signals into luminous representations. This is achieved by making the beams of the CRT scan across the phosphors in order to make them glow and recreate an image.

The scanning beams represent very fine circular spots which traverse across the tube face in a series of fine lines. Since the spots are moving continually, they form an analogue illumination of the tube face. The set of lines covering one complete scan of the tube face is known as a *frame* or *raster*. It will be recalled that, in the TV systems, the set of lines is divided into two half frames or fields. Therefore if one frame consists of 625 lines, then one field would be formed from 312 lines. Again, in the interests of limiting bandwidth, the lines from adjacent fields are scanned in between or interlaced with those of the previous field. In order to avoid a flicker vision effect, frames have to be repeated at least 25 times per second so that 50 interlaced fields per second are used for TV systems. As a useful comparison, it can be shown that this colour TV signal can be digitised to 8-bit resolution at a bit rate of 216 Mbit/s (27MByte/s).

Since most VDUs are viewed from a very close distance, the flicker effect becomes more obvious. Interlaced scanning is therefore replaced by sequential or progressive scanning where successive lines are processed and displayed in their natural order. The VDU often operates at frame and line rates that are linked to the local TV standards. For example, in the USA, the line structure consists of 525 lines at a rate of 15.75kHz, with a field rate of 60Hz. By comparison, the equivalent European line structure consists of 625 lines at a rate of 15.625kHz with a field rate of 50Hz. Since the frame rate effectively represents a new screen image every 16.66ms or 20ms, this is often referred to as the CRT refresh rate.

The persistence of vision of the human eye integrates the light outputs from the three colour phosphors to produce the equivalent coloured images. The slots in the shadow mask tend to make the beam spot rectangular and often for computing purposes, each pixel is considered to be square. The screen resolution for Graphics display is then described in terms of horizontal pixels times vertical pixels. By comparison, the resolution for text display purposes is usually described in terms of characters per row times the number of rows per frame or page. Monitor displays are often characterised by the video standards of IBM personal computers (PCs), of which there is a confusing array. Table 7.1 and its *Notes* give an indication of the variation.

Table 7.1   PC monitor displays

Format	Resolution	Horizontal scan freq.	Vertical scan freq.	Video Signal	Colours
HGA	720 x 348	31.46kHz	60/70Hz	TTL	Mono
MDA	720 x 350	18.43kHz	50Hz	TTL	Mono
CGA	640 x 200	15.75kHz	60Hz	TTL	2 of 16
EGA	640 x 350	21.85kHz	60Hz	TTL	16 of 64
VGA	640 x 480	31 50kHz	60Hz	Analogue	16 of 64
SVGA	800 x 600	48.00kHz	72Hz	Analogue	256 of 256,000
XGA	1,024 x 768	35.52kHz	43Hz	Analogue	256 of 256,000

Notes

HGA (Hercules Graphics Adapter). Combines the graphics facilities of the EGA but in monochrome, with the high quality text of MDA.
MDA (Monochrome Display Adapter). Used for text at 80 characters per row and 25 rows per frame.

CGA (Colour Graphics Adapter). Has two levels of graphics resolution, either 320 *x* 200 pixels in four colours, or a high resolution of 640 *x* 200 in monochrome.

EGA (Enhanced Graphics Adapter). Has graphics resolutions of 640 *x* 350, 640 *x* 200 or 320 *x* 200, plus text capabilities or either 40 *x* 25, or 80 *x* 25.

VGA (Video Graphics Adapter). Provides a wide range of modes for both text and graphics, from 40 *x* 25 to 80 *x* 25 characters and 320 *x* 200 to 640 *x* 480 pixels. At a resolution of 320 *x* 200 it is capable displaying 256 colours.

SVGA (Super Video Graphics Adapter). A high resolution version of VGA. One version of this can display 1,280 *x* 1,024 pixels with a total colour gamut more than 16.7 million (2^{24}).

XGA (Extended Graphics Adapter). A VGA compatible system with the higher resolution of 1,024 *x* 768 pixels which can simultaneously display 256 colours.

MCGA (Multi-Colour Graphics Adapter). Not shown in table. Provides two levels of graphics resolution, either 640 *x* 480 with two colours or 320 *x* 200 with 256 colours, plus text with 80 *x* 30 characters in 16 colours.

In addition there are multi-sync autoscanning, microprocessor controlled VDUs that adapt to suit the output from a computer. The systems that offer lower resolution than VGA are obsolete now, and MCGA is rarely found.

The computer monitor

Reference to Figure 7.3 shows that the monitor has many of the characteristics of the TV receiver. The electron beams are driven horizontally and vertically across the tube face due to the magnetic effects of the sawtooth currents provided by the line and frame scan output amplifiers. The loads for these two stages is provided by the scan coil impedances which has a profound effect on the voltage swings at these two outputs. The power required to obtain full beam deflection can be achieved either by low current and high voltage, high current and low voltage or any of an almost infinite intermediate range of values. For this reason, the scan coil impedances typically vary over the following range: *horizontal*, 0.3mH in series with 0.5Ω, to 1.9mH in series with 2.4Ω; *vertical*, 2mH in series with 1.7Ω, to 110mH in series with 57Ω.

Assuming typical line and frame frequencies of 15.625kHz and 50Hz respectively, calculation of the impedances show that in the case of the line stage, the impedance is almost entirely inductive, whilst for the lower frame frequency the impedance is very largely resistive. The line output stage load appears as an almost perfect integrator so that a sawtooth current can be produced by the pulsed voltage waveform. The line output power amplifier is therefore simply switched on and off at the appropriate time. The load for the frame output stage is

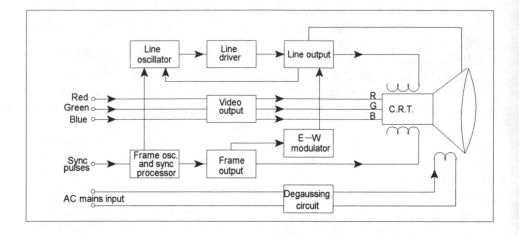

Figure 7.3   Block diagram of CRT display system

very much like that of a low frequency audio amplifier and any small waveform distortion produced by the reactive element can easily be eliminated by using the action of negative feedback.

The displayed raster on the larger wide angle deflection tubes becomes distorted in a pin cushion fashion. This occurs because whilst equal changes of scan current produce equal changes of beam deflection angle, the beam deflection centres do not coincide with the tube face radius. This is usually countered by allowing a small current from each of the horizontal and vertical scan generators to be fed into the other stage. Generally, the vertical or North-South error is small and can be corrected by adding permanent magnets to the scan coil yoke assembly. The larger East-West error is usually corrected by using a diode modulator to correct the line waveform as indicated in Figure 7.3.

In a similar way, the line scan sawtooth waveform becomes slightly distorted and in the wider tubes this is corrected by adding a small capacitor in series with the line scan coils. This is referred to as the S-correction capacitor.

EHT voltage generator

Because of the rapid switching action at the line output stage amplifier, a very large back emf is generated across the windings of the line output transformer (LOPT). Much of this energy can be recovered and used to good effect by including additional windings and rectifiers to provide voltage supplies for other stages of the monitor. The EHT generation is one example of this energy saving process. The typical voltage that can be recovered using a single extra winding is in the order of 8kV, enough to drive a monochrome CRT. This can be multiplied by a factor of about three using a series of diode/capacitor networks to obtain the 25kV needed for the colour CRT.

Degaussing circuit

Because CRTs are sensitive to external magnetic fields, including that of the Earth, they are provided with an internal shield. Over a period of time, this and the shadow mask can develop a degree of permanent magnetism that can create purity problems. Therefore, to provide a demagnetising effect, a circuit that operates at every power-up is included. This consists of a network of a series and a parallel ptc thermistor, to drive a current through coils mounted externally on the tube bulb. This circuit is powered from an ac mains supply. Typically this provides a magneto-motive-force (mmf) of 500At (5A in a 100 turn coil) at switch on, falling to 0.2At (2mA) after about 3 minutes.

Line oscillator, driver and output stage

This circuit as a whole, functions in the manner of a phased locked loop but is more commonly described as a *flywheel synchronisation circuit*. Sampled pulses from the output stage are fed back to the oscillator where their timing is compared with the system synchronising (sync) pulses. Any error then generates a correcting influence to maintain stable operation. Commonly the oscillator and driver stage are contained within an IC with only the output stage using discrete components. The oscillator waveform is commonly square and its mark-to-space ratio is modified by feedback to obtain the necessary switching action at the output stage. The power amplifier and LOPT are normally screened/encapsulated to minimise X-ray radiation and to isolate the very high voltages that are present. The number of scanning lines per raster is given by the ratio of the horizontal to vertical timebase frequencies (always an integer value). Thus both stages are synchronised to the same common composite sequence of sync pulses.

Frame oscillator and output stage

Because this stage operates at lower frequencies, it is much less stressed than the line output stage. The design typically follows audio amplification principles but uses extensive feedback to ensure that the output deflection current is a perfect sawtooth. The whole circuit is commonly contained within a single IC.

Synchronising

Reference to Figure 7.2 shows that the composite video signal has two components, image information and synchronising pulses. In addition, provision needs to be made for the system circuits to change over between these two types of processing and to allow for both horizontal (line) and vertical (frame) beam retrace. Because the video signal is blanked to black level, these periods are referred to as horizontal (HBI) and vertical blanking intervals (VBI) respectively. The two short periods on either side of the line sync pulse are known as the front and back porches (resting periods). Because the vertical retrace takes several line periods to prepare for the start of the next frame, not all the raster lines are available for the image display. The proliferation of raster standards mentioned above, continues with the variation of sync pulse formats. Figure 7.2 shows that the line sync pulse is negative going (–) with respect to black level. However there are many different systems is use. Both line and frame sync pulses may be positive (+) or (–) and in the former case are often carried on a separate signal line. For analogue R, G, B signals, the sync pulses may be added to the green signal (sync on green) in the manner indicated by Figure 7.2.

Video drives and amplifiers

The bandwidth of this section of the monitor must be high enough to ensure that the display is presented with a high degree of resolution. This can be shown by calculating the bandwidth required at the CRT. Taking the medium resolution of VGA images as an example, each frame is represented by $640 \times 480 = 307,200$ pixels. Since the maximum frequency occurs when succeeding pixels are repeatedly switched on and off, this represents $\frac{1}{2}(640 \times 480)$ cycles per frame. With a refresh rate of 60Hz, the upper frequency becomes $153,600 \times 60Hz = 9.216MHz$. This value rises to 17.28MHz for SVGA.

For colour displays, this bandwidth requirement exists for each of the red, green and blue stages. Video amplifiers are characterised by the use of low values of load resistance with high supply voltages. Between the amplifier output and the CRT input, there exists a significant stray capacitance and this produces a shunting effect which reduces the effective gain. The upper –3dB frequency occurs at the point where the load resistance and stray capacity reactance values are equal.

Two common techniques that are used in order to drive this point as high as possible, include the addition of an inductor in series with the load and the application of selective negative feedback (NFB). The inductor which is known as a *peaking choke* has a reactance that rises with frequency and so cancels the shunting effect of the stray capacitance. Using a NFB system whose feedback factor falls with rising frequency, allows the amplifier gain to rise, again compensating for the stray capacitance.

The high voltage supply (typically at least 100V) is necessary to provide the high level of signal needed to fully drive the CRT. Typically the video amplifier is based on a three channel (R, G, B,) integrated circuit (IC) which is in turn based on the operational amplifier driving a compound output stage. The circuit then

108

uses overall feedback in order to obtain the necessary gain and bandwidth. Because the frequency response has to extend down to dc, the upper −3dB (f_u) point is also the bandwidth. Because of the high rates of signal level change and the large amplitudes, the amplifier rise time (t_d) or the slew rate are the more important parameters. As would be expected, these three parameters are related: $t_d = 0.35/f_u$ to $0.45/f_u$ and Slew rate = $2\,f_u(V_{out.max})$.

A typical IC may have the following open loop parameters:

      Slew rate > 2000V/µs

      Bandwidth > 15MHz

      Rise time < 50ns, and

      Open loop gain >50dB.

Under operational conditions such an amplifier might have the following parameters:

      Closed loop gain >20dB

      Signal power output per channel <3W

      Peak to peak voltage swing <180V

      Peak output current per channel <18mA.

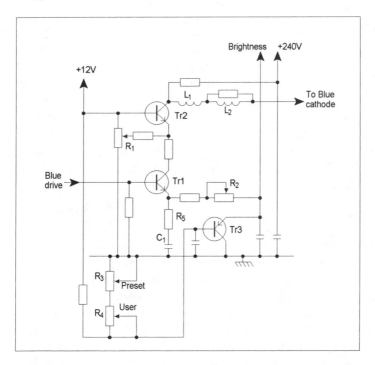

Figure 7.4   Discrete component video amplifier for a colour monitor

The discrete component circuit shown in Figure 7.4 indicates the principles involved in video amplification. The complete circuit consists of three identical parallel amplifiers, one for each channel (R, G, B). The gain and bandwidth is obtained by using a common emitter, common base compound amplifier (Tr1 and Tr2) , with peaking chokes (L_1 and L_2) and NFB via R_5 and C_1. The brightness control for all three stages is provided by Tr3, R_3 and R_4.

Each stage contains two preset controls R_1 and R_2. These are adjusted as follows to set the beam cut-off point and peak beam current respectively. R_1 sets the black level or point at which each gun starts to conduct and R_2 is used to set peak white. Finally the brightness and contrast controls are set for equal brightness changes between each step of the grey scale test signal. The dark areas of the image are then viewed for signs of colouration and R_1 is adjusted accordingly. Similarly, the bright areas are observed and R_2 adjusted as necessary. This operation should leave an equal graduation grey scale pattern without false colouration. For CRTs using colour-difference signal drives, this adjustment sequence is subtly different.

The colour gamut or range of colours

When the three guns are driven by analogue signals, there is virtually an infinity of different colours available for display. However, with the guns under digital control, there is a significant difference. With three primary colours, red, green and blue, each capable of being on or off, only $8 = 2^3$ (including black) are available. If each primary colour is allowed to exist at half amplitude, the three colours and three possible amplitudes provide for $3^3 = 27$ different colours. For the SVGA format using colour signals with 8-bit resolution for each channel, there are $2^8 \times 2^8 \times 2^8 = 2^{24} = 16.777216$ million different colours or shades available. It should be noted that for any given screen format, not all the colours may be available for simultaneous use.

Screen memory requirements

Unlike television, where the screen data is continually being transmitted, the computer monitor receives its information via the system memory, the capacity of which controls the way in which screen images are displayed. For example, in text mode, only a small part of the screen is illuminated at any one time. The memory size needed to support this operation then depends on the text density. For example, in CGA mode with 25 rows each of 80 characters, the memory needs to hold up to 2,000 different ASCII characters. For display purposes, it is also necessary to be able to describe the character background, thus each text symbol requires 2 bytes. The memory requirements in text mode can be calculated from, Characters per row x Rows per screen x 2.

Thus a CGA frame or page represents nearly 4Kbytes of memory. Since any given memory has rather greater capacity than this, several pages of text may be simultaneously held in the screen memory. For the graphics mode, the whole screen may be simultaneously illuminated thus requiring much greater memory capacity. (Often described as *all points addressable,* or APA). This obviously depends upon the screen resolution in horizontal and vertical pixels and the number of bits necessary to describe the available colours.

For colour representation the number of bits required can be calculated as follows:

1 bit, 2 colours, (monochrome)
2 bits, 4 colours
3 bits, 8 colours
4 bits, 16 colours
n bits, 2^n colours.

For the CGA format, using 320×200 pixel resolution with 4 colours, requires $320 \times 200 \times 2 = 128$Kbit per screen. At a resolution of 640×400 with 2 colours, this becomes, $640 \times 400 \times 1 = 256$Kbit per screen.

By comparison, the XGA format with 1024×768 pixels and 256 (8 bits) colours requires $1024 \times 768 \times 8 = 6.291456$Mbit per screen.

Alphanumeric and graphics character symbols

The 8 bit codes used for processing character symbols within a computer are only indirectly used for creating the equivalent screen display symbol. Each unique character shape is represented on screen by a cell or matrix array of rows and columns of dots with each dot being capable of illumination or not. The size of each cell varies from 15×12 for high definition systems, through 8×8, to 10×6, which is used extensively in general purpose processing. Quite obviously, the finest detail or resolution that can be displayed is controlled by the pixel dimensions which are in turn, a feature of the line scan width and the CRT design.

Figure 7.5(a) indicates the way in which each dot of the matrix is organised for a 10×6 character cell. Since column 6 and row 10 is left blank to define an inter-character border and rows 8 and 9 are provided for symbols with descenders (i.e. g, j, y, etc), the major area of the character cell occupies 7×5 dots. In cases where interlaced line scanning is employed, each character cell is generated during each field so that the cell contains a total of 20 rows.

1	1	1	1	1	0
1	0	0	0	0	0
1	0	0	0	0	0
1	1	1	1	1	0
0	0	0	0	1	0
0	0	0	0	1	0
1	1	1	1	1	0
0	0	0	0	0	0
0	0	0	0	0	0
0	0	0	0	0	0

b_1	b_2
b_3	b_4
b_5	b_6

(b)

(a)

Figure 7.5   Character display cell: (a) alphanumerics, (b) graphics

The basic principle of the character generator is shown in Figure 7.6. The ROM holds all the dot matrix patterns stored in cells in the manner shown in Figure 7.5(a), for all the symbols in the set.

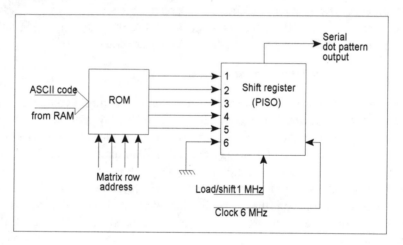

Figure 7.6   Basic principles of character generator

Table 7.2 ISO-7 character set

Col Row	0	1	2	2a	3	3a	4	5	6	6a	7	7a
0	NUL[1]	DLE[1]					@	P			p	
1	Alphan Red	Graphics Red	!		1		A	Q	a		q	
2	Alphan Green	Graphics Green	"		2		B	R	b		r	
3	Alphan Yellow	Graphics Yellow	£		3		C	S	c		s	
4	Alphan Blue	Graphics Blue	$		4		D	T	d		t	
5	Alphan Magenta	Graphics Magenta	%		5		E	U	e		u	
6	Alphan Cyan	Graphics Cyan	&		6		F	V	f		v	
7	Alphan White	Graphics White[2]	'		7		G	W	g		w	
8	Flash	Conceal Display	(		8		H	X	h		x	
9	Steady[2]	Contiguous Graphics[2]	)		9		I	Y	i		y	
10	End Box[2]	Separated Graphics	*		:		J	Z	j		z	
11	Start Box	ESC[1]	+		;		K	←	k		¼	
12	Normal[2] Height	Black[2] Background	,		<		L	½	l		‖	
13	Double Height	New Background	−		=		M	→	m		¾	
14	SO[1]	Hold Graphics	.		>		N	↑	n		÷	
15	SI[1]	Release[2] Graphics	/		?		O	#	o			

Notes:
[1] These control characters are reserved for compatibility with other data codes.
[2] These control characters are presumed before each row begins.
Codes may be referred to by their column and row, e.g. 2/5 refers to % sign.
Black represents display colour, white represents background. Rectangle shows size of character outline.

Table 7.3   IBM PC-8 character set

NUL 0	► 16	SP 32	0 48	@ 64	P 80	` 96	p 112	Ç 128	É 144	á 160	▓ 176	└ 192	╨ 208	α 224	≡ 240
☺ 1	◄ 17	! 33	1 49	A 65	Q 81	a 97	q 113	ü 129	æ 145	í 161	▒ 177	┴ 193	╤ 209	β 225	± 241
● 2	↕ 18	" 34	2 50	B 66	R 82	b 98	r 114	é 130	Æ 146	ó 162	▓ 178	┬ 194	╥ 210	Γ 226	≥ 242
♥ 3	‼ 19	# 35	3 51	C 67	S 83	c 99	s 115	â 131	ô 147	ú 163	│ 179	├ 195	╙ 211	π 227	≤ 243
♦ 4	¶ 20	$ 36	4 52	D 68	T 84	d 100	t 116	ä 132	ö 148	ñ 164	┤ 180	─ 196	╘ 212	Σ 228	⌠ 244
♣ 5	§ 21	% 37	5 53	E 69	U 85	e 101	u 117	à 133	ò 149	Ñ 165	╡ 181	┼ 197	╒ 213	σ 229	⌡ 245
♠ 6	▬ 22	& 38	6 54	F 70	V 86	f 102	v 118	å 134	û 150	ª 166	╢ 182	╞ 198	╓ 214	μ 230	÷ 246
• 7	↨ 23	' 39	7 55	G 71	W 87	g 103	w 119	ç 135	ù 151	º 167	╖ 183	╟ 199	╫ 215	τ 231	≈ 247
◘ 8	↑ 24	(40	8 56	H 72	X 88	h 104	x 120	ê 136	ÿ 152	¿ 168	╕ 184	╚ 200	╪ 216	Φ 232	° 248
○ 9	↓ 25	) 41	9 57	I 73	Y 89	i 105	y 121	ë 137	Ö 153	⌐ 169	╣ 185	╔ 201	┘ 217	Θ 233	· 249
◙ 10	→ 26	* 42	: 58	J 74	Z 90	j 106	z 122	è 138	Ü 154	¬ 170	║ 186	╩ 202	┌ 218	Ω 234	· 250
♂ 11	← 27	+ 43	; 59	K 75	[91	k 107	{ 123	ï 139	¢ 155	½ 171	╗ 187	╦ 203	█ 219	δ 235	√ 251
♀ 12	⌐ 28	, 44	< 60	L 76	\ 92	l 108	| 124	î 140	£ 156	¼ 172	╝ 188	╠ 204	▄ 220	∞ 236	ⁿ 252
♪ 13	↔ 29	- 45	= 61	M 77	] 93	m 109	} 125	ì 141	¥ 157	¡ 173	╜ 189	═ 205	▌ 221	φ 237	² 253
♫ 14	▲ 30	. 46	> 62	N 78	^ 94	n 110	~ 126	Ä 142	₧ 158	« 174	╛ 190	╬ 206	▐ 222	ε 238	■ 254
☼ 15	▼ 31	/ 47	? 63	O 79	_ 95	o 111	⌂ 127	Å 143	ƒ 159	» 175	┐ 191	⊥ 207	▀ 223	∩ 239	SP 255

The ROM therefore acts as a look-up table to aid the conversion between the two character formats. For example, an ASCII or ISO-7 code character stored in the RAM is input to a 7 – 128 line decoder to form the cell address in the ROM which holds that particular character shape. A 4-bit row address code then identifies the particular row holding the necessary dot pattern. These 5 bits are then read out in parallel and converted into serial format by the shift register to be clocked out by the 6 MHz dot clock to form the bright-up video signal. The character generator may also be used to provide the flashing cursor symbol. Character generators are also available for the PC-8 or the EBCDIC (extended binary coded decimal interchange code). Since these are both 8 bit codes, the ROM address selector is an 8 – 256 line decoder coupled to a larger ROM.

This can be shown as follows:

Number of bits = Number of characters x number of rows x number of columns.

The 128-character set using 9×5 matrix requires 5,760 bits; the 256-character set using the same matrix requires 11,520 bits.

Table 7.2 illustrates the ISO-7 character set, and Table 7.3 shows the PC-8 8-bit set that is used on printers which are intended to work with the IBM PC type of computer. Most printers are capable of using any of a number of character sets.

Colour decoding of the dot pattern

Reference to Table 7.2 (ISO-7 code) shows that b_4, b_6 and b_7 are always set to logic 0 when a coloured display is needed. At the same time, b_5 acts as a switch between alphanumeric and graphics coloured symbols. Whilst Figure 7.7 shows the principle of this function, in practice, its circuitry will normally be buried within an IC.

The output from gate G_4 acts as a clock signal for the three D-type flip-flops, to gate the signals from b_1, b_2 and b_3 to the R, G, B CRT output drives along with the dot pattern from the character generator. The line sync pulse acts as a reset to the flip-flops to ensure that each scan line starts up in the same mode.

Graphics displays

For simple graphic symbols the 10×6 character cell is sub-divided into a 3×2 cell in the manner shown in Figure 7.5(b). However, since the original 10-line cell does not divide equally into three sections, the middle one is allocated 4 lines. Since these cells are usually displayed without borders they are described as contiguous graphics. Each of these cells are identified by logic 1 for illumination and logic 0 for black. The general principles of the control of alphanumeric display is adopted for graphics. For those applications where the raggedness of such graphics displays is unacceptable, it is possible to use an inter-line comparator to detect where the large steps occur. The controller then inserts a quarter area block to produce character rounding.

For high resolution graphics such as CAD/CAM, each pixel may be bit mapped or *all points addressable* (APA). The following calculation indicates how the screen display memory size increases for such an application. Consider the CGA format in low resolution mode (320×200 pixels): with 16 colours the memory

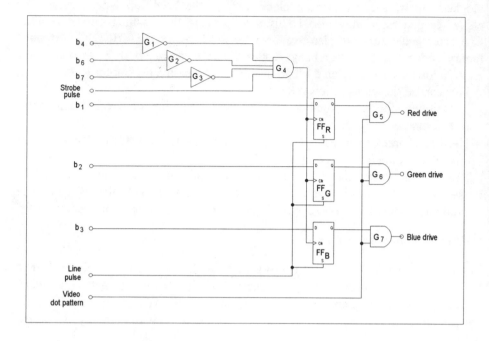

Figure 7.7   Colour decoding logic

needs a capacity of 320 x 200 x 4 bits = 256Kbits. Using the same memory for high resolution of 640 x 200 pixels allows for only 256000/(640 x 200) = 2 bits. Thus CGA only provides for 4 colours in the high resolution mode.

Cathode ray tube controller (CRTC)

The CRTC acts as a system controller for the VDU. In general, it sits between the display system, the processor and any I/O (input/output) device that connects the computer to the outside world. So called *intelligent* VDUs/terminals have a dedicated microprocessor embedded in the CRTC. This operates under stored program control from data stored in a ROM and responds to interrupt signals from I/O devices. The basic functions of this device can be summarised as follows:

1.   Generate line and frame timebase synchronising signals for the display device.
2   Driven from control codes stored in RAM, the CRTC can change the colour display and generate a cursor signal if needed.
3.   Provide Read/Write control signals and addresses for the RAM. The required ASCII or other equivalent code symbols are stored in the RAM

116

during the frame flyback or vertical blanking period. The corresponding dot codes are output from the RAM during the active line scan period.

4. When the ASCII or other code signals are provided from the RAM to the character generator ROM, the CRTC provides the row addresses, the load/shift signal and the clock pulse needed to generate the dot pattern for each character.

Figure 7.8 shows the basic principle behind the interface between the main processor and the CRTC. This employs *direct memory access* (DMA) to obtain a fast movement of data between the RAM and the character generator by using block transfer. When an I/O device generates an interrupt signal to the controller (DMAC), this in turn generates a hold signal to the microprocessor to disconnect it from the bus system. Data is then transferred from the RAM into one of the two buffer stores via a switching arrangement and at the same time, data is being read out from the other store. At an appropriate point in time, the switches change over and the other buffer store is loaded.

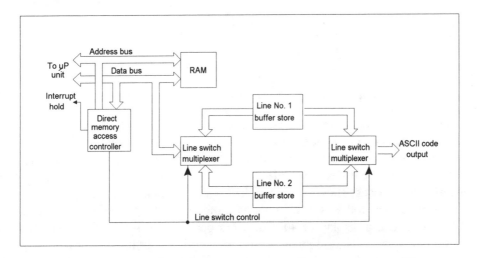

Figure 7.8   Interface to CRT controller

Because loading is very much faster than read-out, the processor is only interrupted for a relatively brief period of time. The modern CRTC IC may contain an embedded microprocessor to impart a degree of intelligence to the interface and the general principle of this is shown in Figure 7.9.

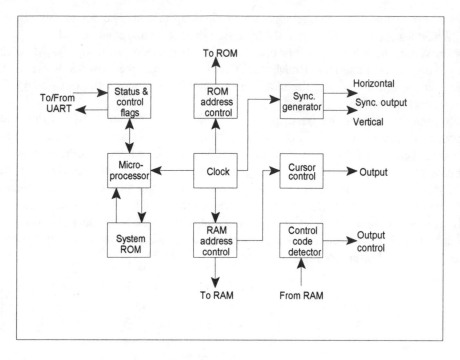

Figure 7.9   Basic principles of intelligent CRTC

Whilst the central element, the clock or timing circuit, is used to ensure synchronism, the microprocessor is used to handle the status and interrupt signals. The clock circuit is much more complex than inferred by Figure 7.9. The major aim being to synchronise the video dot pattern to the line and frame timebase signals. This process shown in Figure 7.10, can be explained by using an example for a simple system that is compatible with the world standard teletext system.

This provides a display of 40 characters per line, 24 rows of text per frame, 10 x 6 character cells, using interlaced scanning for either the 525 or 625 line TV systems. Pages or frames from this database can be displayed on many personal computers. A period of 40μs is allocated for text display on every line, thus with 1 character per μs and 6 dots per character, requires a crystal controlled dot clock oscillator running at 6MHz. Division by 6 provides the Row address count for the ISO-7 codes held in RAM, whilst a further division by 64 provides an output at line timebase rate. Division of the line rate by 10 provides the Row address for the character generator ROM pattern and division by 24 identifies the text row of the display. The 1MHz signal is used to provide a load/shift character command to the output shift register and the 6MHz signal controls the output dot pattern rate. The RAM is loaded whilst the read/write line is held low during the vertical blanking interval, and read out during the active line period.

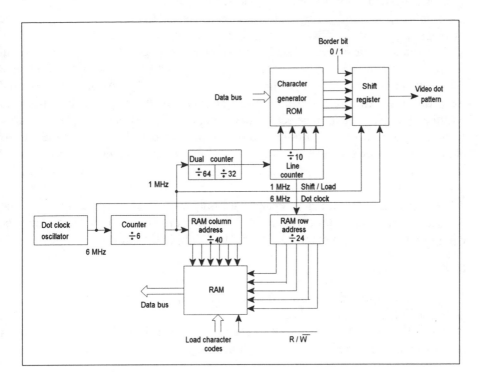

Figure 7.10   Timing network for CRTC

An alternative division, by 32 instead of 64, ensures that each video line dot pattern is read out twice in succession, thus producing double height characters.

The universal asynchronous receiver transmitter (UART)

This element forms one of the common interfaces between the computer and VDU, and the outside world. Basically it converts serial data input into parallel format for computer processing and then performs the complementary operation to provide an output. The device is also variously known as a USART (universal synchronous asynchronous receiver transmitter) or USRT (universal synchronous receiver transmitter), both of which can perform either synchronous or asynchronous data transfers. The synchronous mode often provides the highest date transfer rate. An alternative title is the ACIA (asynchronous communications interface adapter), all of which are 40-pin LSI chips. By comparison, a 48-pin device known as a DUSCC (dual universal serial communications controller) provides for the simultaneous communications over two independent full duplex channels in a single chip.

As can be seen from Figure 7.11, the UART has three basic sections, receive, transmit and control. The latter receives its instructions either from the microprocessor or the CRTC and generates a number of flag or status signals. This diagram indicates the degree of symmetry between the input and output circuits. As a controller of serial data transfers, it must be able to insert and delete the start and stop bits used to delineate each data byte. A *start* bit is represented by a transition from high to low and a *stop* bit by the opposite transition. In addition, it must be able to select the data rate, carry out either odd or even parity tests, set the number of stop bits and the number of data bits or word length as called for by system design. This may require even, odd or no parity; word lengths of 5, 6, 7, or 8 bits; and either 1, 1½ or 2 stop bits. (1 being reserved for 5-bit codes).

Most UARTs are similar in design and only vary in the special features that make then adaptable to a particular microprocessor. For example, some devices can provide the modem control lines of CTS, RTS, DSR and DTR.

Control and status lines

The following provides a brief description of the actions that result from the different signal levels on various lines.

The EPE and PI lines are used to set the system parity. The PI line disables parity checking altogether whilst a high or a low on EPE sets even or odd parity respectively.

The SBS line enables the number of stop bits to be set at 1 or 2.

The CL1 and CL2 lines are used to set the data word length.

A low on the TBRL line allows the transmitter buffer register to be loaded in the transmission mode.

A high on the RRD line disables the tri-state buffers at the output in the receive mode. A low on this line enables the parallel outputs.

A high on the PE line signifies a parity check failure.

A high on the TRE line signals that the transmitter register is empty and the complete multiplexed character stream has been transmitted.

A high on the DR line indicates that a character has been transferred to the receiver buffer register without detectable errors.

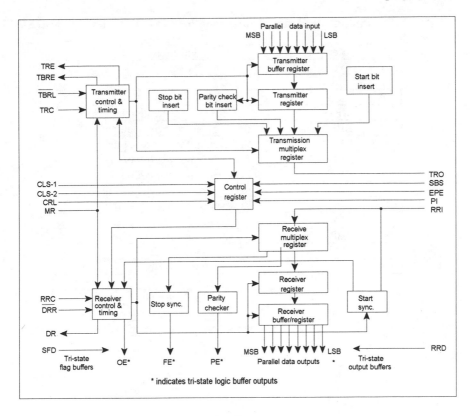

Figure 7.11   Block diagram of typical UART

Exercise 7.1

Observe the CRT mountings, scan coil assembly, convergence assembly and degaussing coils. Study the safety procedures for changing the CRT and if possible carry out this operation.

Exercise 7.2

Carry out the degaussing operation and then make the necessary purity adjustments. Once the purity has been correctly set, the effect of the Earth's weak magnetic field can be seen by turning the monitor upside down.

Table 7.4   Pin outs and signal functions for UART RS-6402 (Figure 7.11)

Pin	Key	Function
1	Vcc	+v supply
2	N/C	
3	Gnd	−v supply
4	RRD	Receiver register disable
5	RBR8	Receiver buffer register output
6	RBR7	Receiver buffer register output
7	RBR6	Receiver buffer register output
8	RBR5	Receiver buffer register output
9	RBR4	Receiver buffer register output
10	RBR3	Receiver buffer register output
11	RBR2	Receiver buffer register output
12	RBR1	Receiver buffer register output
13	PE	Parity error
14	FE	Framing error
15	OE	Over-run error
16	SFD	Status flags disable
17	RRC	Receiver register clock
18	DRR	Data received reset — active low
19	DR	Data received
20	RRI	Receiver register input
21	MR	Master reset
22	TBRE	Transmitter buffer register empty
23	TBRL	Transmitter buffer register load — active low
24	TRE	Transmitter register empty
25	TRO	Transmitter register output
26	TBR1	Transmitter buffer register inputs
27	TBR2	Transmitter buffer register inputs
28	TBR3	Transmitter buffer register inputs
29	TBR4	Transmitter buffer register inputs
30	TBR5	Transmitter buffer register inputs
31	TBR6	Transmitter buffer register inputs
32	TBR7	Transmitter buffer register inputs
33	TBR8	Transmitter buffer register inputs
34	CRL	Control register load
35	PI	Parity inhibit
36	SBS	Stop bit select
37	CLS1	Character length select
38	CLS2	Character length select
39	EPE	Even parity enable
40	TRC	Transmitter register clock

Exercise 7.3

Carry out the convergence operation to ensure that beam landing errors are minimised over the whole tube face area. Note that this operation may not be possible on certain types of CRTs. If it is possible, it is carried out in two stages; the static adjustment affecting the central area of the screen and dynamic adjustment for the rest of the screen and general overall effect.

Exercise 7.4

Carry out picture shift adjustments and set the EHT voltage. Note that it is important to be guided by manufacturers service instructions.

The monitor programs listed in Appendix 1 and referred to in this chapter are provided by courtesy of Mr K. P. Taylor, 15 Lindsay Road, Horfield, Bristol, BS7 9NP, who can supply a much more extensive suite of such facilities.

Test Questions

1 (a) Calculate the memory capacity required for a system capable of displaying graphics in 16 colours with a screen resolution of 640 x 480 pixels.

(b Calculate the memory capacity required for a text display using 80 characters per line and 25 lines per page.

(c) How many pages of text in (b) could be stored in the memory of (a)?

2 (a) Name the three primary and three secondary colours used for a CRT colour display.

(b) State the colours obtained from a standard colour bar test waveform when the Green gun is inoperative. (Colour bars — White, Yellow, Cyan, Green, Magenta, Red, Blue, Black.)

3. Describe four important safety precautions to be observed when changing a CRT.

4. Explain the functions performed by the CRT rimband.

5. Calculate the total bit rate of a colour video signal with bandwidths for Red, Green and Blue components extending to 8MHz and using 8 bits per sample. What is the bandwidth of the total digital signal?

8   Logic systems and faultfinding

Syllabus references: D06 – 6.1,D06 – 6.2, D06 – 6.3, 14.1, 14.2

Servicing instruments and methods

In some respects, servicing logic circuitry can be simpler than work on analogue circuits of comparable size. All digital signals are voltages at one of two levels, and there is no problem of identifying minor changes of waveshape which so often cause trouble in linear circuits. In addition, the specifications that have to be met by a microprocessor circuit can be expressed in less ambiguous terms than those that have to be used for analogue circuits. You don't, for example, have to worry about harmonic distortion or intermodulation, and parasitic oscillation is rare, though not impossible.

That said, logic and microprocessor circuits bring their own particular head-aches, the worst of which are the relative timing of voltage changes and the difficulty of displaying signals. As the timing diagrams in Chapter 11 will demonstrate, the actions of any microprocessor circuit depend on strict timing being maintained, and conventional equipment which serves well for the analysis of analogue circuits is of little use in working with microprocessor circuits. The problem is compounded by the fast clock rates that have to be used for many types of microprocessors.

You may, for example, be looking for a coincidence of two pulses with a 33MHz clock pulse, with the problem that the coinciding pulses happen only when a particular action is taking place. This action may be completely masked by many others on the same lines, and it is in this respect that the conventional oscilloscope is least useful. Oscilloscopes as used in analogue circuits are intended to display repetitive waveforms, and are not particularly useful for displaying a waveform which once in 300 cycles shows a different pattern. A fast conventional oscilloscope is useful for checking clock pulse rise and fall times, and for a few other measurements, but for anything that involves bus actions a good storage oscilloscope is needed.

In addition, some more specialised equipment will be necessary if anything other than fairly simple work is to be contemplated. Most of this work is likely to be on machine control circuits and the larger types of computers. Small computers do not offer sufficient profit margin in repair work to justify much diagnostic equipment. After all, there's not much point in carrying out a £200 repair on a machine which is being discounted in the shops to £250! We shall start this Chapter looking at some specialised instruments for digital circuit work.

Current tracer

Work on printed circuit boards makes it difficult to trace currents, since it would be unacceptable to split a track to check the current flowing through it. A current-tracer (also called current checker) is a small hand-held device that makes it possible to detect the presence and direction of current flow on a PCB track, and also to indicate, roughly, its size.

The principles of one type are illustrated in Figure 8.1. Two probes, set a fixed distance apart, are pressed on the track, and the tiny voltage difference that is produced by the current flow is amplified by balanced operational amplifiers so

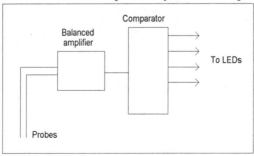

Figure 8.1   Principles of a simple current tracer

as to operate LED indicators. Typically, three LEDs are used to indicate currents of 10, 50 and 100mA on 1mm track width, or 20, 100 and 200mA on 2mm track width. A fourth LED is used to indicate reverse polarity, or to indicate open-circuit track.

An alternative method of tracing current uses a Hall-effect probe which does not rely on sensing voltage, using instead the magnetic field around the track. This type of tracer, such as the ToneOhm, manufactured by Polar Instruments, will detect partial and complete short circuits and, with practice, can be used to find the position of an open-circuit.

A current tracer, like a voltmeter, is a useful first-test instrument, as it will assist in finding any problems that cause a drastic change in the current flowing along a track. This can help to pin-point damaged tracks (a hazard particularly in flexible circuit strips) or failure of components that draw large currents, such as printer-head driver transistors. One point to note, however, is that modern PCBs with narrow tracks make the use of any type of current tracer very much more difficult.

Logic probe

A logic probe is a device which uses a small conducting probe to investigate the logic state of a single line. The state of the line is indicated by LEDs, which will indicate high, low or pulsing signals on the line. The probe is of very high impedance, so that the loading on the line is negligible. Figure 8.2 shows a block diagram for a typical instrument.

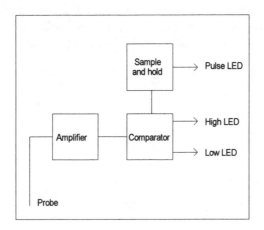

Figure 8.2   A simplified block diagram for a logic probe

The voltage on the track is amplified, and applied to a comparator which will detect high or low state. If the voltage is intermediate, as can be caused by the presence of pulses, the sample and hold circuit will rectify and store the pulse level, activating the third LED.

A typical logic probe can be switched to either TTL or CMOS voltage levels (+5V supply for TTL and 3V–18V for CMOS) and uses coloured LEDs to identify the logic state of the track. Pulses as narrow as 30ns can be detected, and as wide as 500ms. Some types can also indicate the presence of ripple on the power supply, indicating faulty stabilisation. Probes can be battery-powered, or can take their power from the circuit under test, using crocodile clips.

These probes are not costly, around £15 to £50, and are extremely useful for a wide range of work on faults of the simpler type. They will not, obviously, detect problems of mistiming, but such faults are rare if a circuit has been correctly designed in the first place. Most straightforward circuit problems, which are mainly chip faults or open or short circuits, can be discovered by the intelligent use of a logic probe, and since the probe is a pocket-sized instrument it is particularly useful for on-site servicing.

Obviously, the probe, like the voltmeter used in an analogue circuit, has to be used along with some knowledge of the circuit. You cannot expect to gain much from simply probing each line of an unknown circuit. For a circuit about which little is known, though, some probing on the pins of the microprocessor can be very revealing. Since there are a limited number of widely-used microprocessor types, it is possible to carry around a set of pinouts for all the microprocessors that will be encountered.

Starting with the most obvious point, the probe will reveal whether a clock pulse is present or not. Quite a surprising number of defective systems go down with this simple fault: it is even more common if the clock circuits are external. Other very obvious points to look for are a permanent activating voltage on a HALT line, or a permanent interrupt voltage, caused by short circuits. For an intermittently functioning or partly functioning circuit, failure to find pulsing voltages on the higher address lines or on data lines may point to microprocessor or circuit-board faults.

For computers, the description of the fault condition along with knowledge of the service history may be enough to lead to a test of the line that is at fault. The considerable advantage of using logic probes is that they do not interfere with the circuit, are very unlikely to cause problems by their use, and are simple to use. Some 90% of microprocessor system faults are detectable by the use of logic probes, and they should always be the first hardware diagnostic tool that is brought into action against a troublesome circuit.

Logic pulsers

Logic pulsers (or digital pulsers) are the companion device to the logic probe. Since the whole of a microprocessor system is software operated, some lines may never be active unless a suitable section of program happens to be running. In machine control circuits particularly, this piece of program may not run during any test, and some way will have to be found to test the lines for correct action. A typical block diagram is illustrated in Figure 8.3.

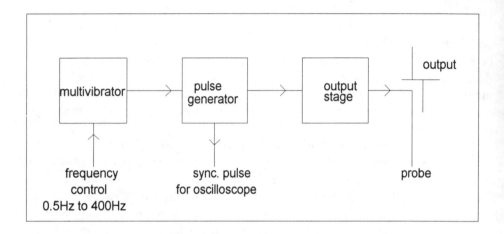

Figure 8.3   A typical block diagram for a logic pulser

A logic pulser, as the name indicates, will pulse a line briefly, almost irrespective of the loading effect of the chips attached to the line. The injected pulse can be detected by the logic probe. This method is particularly useful in tracing the path of a pulse through several gate and flip-flop stages.

The multivibrator generates a square wave whose frequency can be varied in the range 0.5Hz to 400Hz, and this wave is used to trigger a pulse generator which provides positive and negative pulses of about 10μs width. The output stage will float to the voltage of the line that the probe is touching, and can source or sink up to 100mA. This is enough to override any other drivers on the line. A synchronising pulse can be used to trigger an oscilloscope if needed.

The logic pulser is a more specialised device than the logic probe, and it has to be used with more care. It can, however, be very useful, particularly where a diagnostic program is not available, or for testing actions that cannot readily be simulated.

Logic clip

The logic clip is an extension of the logic probe to cover more than one line. As the name suggests, these devices clip over a logic IC, and are available for 14-pin or 16-pin DIL packages. As the block diagram of Figure 8.4 shows, each pin of the IC is connected to a buffer in the logic clip, and this buffer drives an LED indicator. Logic clips are usually available in separate TTL or CMOS versions, though the TTL version is now more common because of the widespread use of 74HC devices.

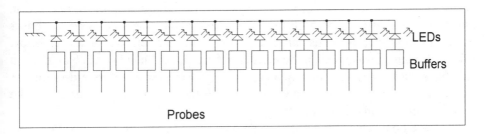

Figure 8.4   Block diagram for a logic clip

The logic clip is particularly useful when the system clock rate can be slowed down, or a logic pulser is being used to supply an input. Because all the states on a single IC chip can be monitored together, any fault in a gate within the chip is fairly easy to find, much easier than the use of a logic probe on all signal pins in turn.

Logic comparator

A logic comparator is a 'bench' version of the logic clip which also uses an LED indicating the state of each line of 16, 32, 40, or more, lines. Connection to the circuit is made through ribbon cable, terminating usually in a clamp which can be placed over an IC. For older microprocessor circuits, the usual clamp is a 40-pin type, and the ends of the ribbon cable must be attached to the set of bus lines for that particular microprocessor. Pre-wired clamps are often available for popular microprocessor types.

The LED is lit for logic high, and unlit for logic zero or the floating state. For a pulsing line, the brightness of each LED is proportional to the duty cycle of the pulses on the line. Most logic monitors have variable threshold voltage control, so that the voltage of transition between logic levels can be selected to eliminate possible spurious levels. The use of a 40-point monitor is much more useful for industrial microprocessors, since only a knowledge of the microprocessor pinout will then be needed. Since port chips are generally in a 40-pin package also, this allows tests on ports, which are often a fruitful source of microprocessor system troubles.

Logic analysers

The logic analyser is an instrument which is designed for much more detailed and searching tests on digital circuits generally and on microprocessor circuits in particular. As we have noted, the conventional oscilloscope is of limited use in microprocessor circuits because of the constantly changing signals on the buses as the microprocessor steps through its program. Storage oscilloscopes allow relative timing of transitions to be examined for a limited number of channels, but suitable triggering is seldom available. Logic probes and monitors are useful for checking logic conditions, but are not helpful if the fault is one that concerns the timing of signals on different lines.

The logic analyser is intended to overcome these problems by allowing a time sample of voltages on many lines to be obtained, stored, and then examined at leisure. Most logic analysers permit two types of display. One is the 'timing diagram' display, also called 'timing domain analysis' in which the various logic levels for each line are displayed in sequence, running from left to right on the output screen of the analyser. A more graphical form of this display can be obtained by connecting a conventional oscilloscope, in which case, the pattern will resemble that which would be obtained from a 16-channel storage oscilloscope. The synchronisation may be from the clock of the microprocessor system, or at independent (and higher) clock rates which are more suited to displaying how signal levels change with time.

The other form of display is word display or 'data domain analysis'. This uses a reading of all the sampled signals at each clock edge, and displays the results as a 'word' for each clock pulse rather than as a waveform. If the display is in binary, then the word will show directly the 0 and 1 levels on the various lines. For many purposes, display of the status word in other forms, such as hex, octal, denary or ASCII, may be appropriate. This display, which gives rise to a list of words as the system operates, is often better suited for work on a system that uses buses, such as any microprocessor system.

The triggering of either type of display may be at a single voltage transition, like the triggering of an oscilloscope, or it may be gated by some preset group of signals, such as an address (a trigger word or event). This allows for detecting problems that arise when one particular address is used, or one particular instruction is executed. One common method is to trigger a display by using a combination of inputs. One of these would be a trigger word which can be set and stored, the other inputs would be trigger signals (qualifiers) which can be taken from the clock (a clock qualifier) or from other inputs (trigger qualifier). You can also use a word search (trace word or event action) through the memory of the analyser to find if a specified word has been stored in the course of an analysis.

Figure 8.5 shows a simplified block diagram of a typical analyser.

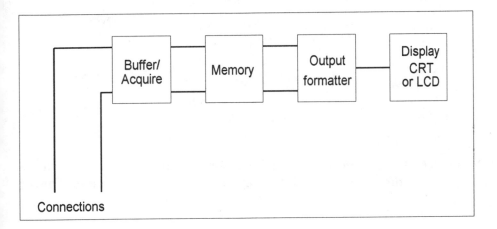

Figure 8.5   Simplified block diagram for a logic analyser

The data acquisition portion of an analyser is illustrated in the block diagram of Figure 8.6. The logic levels at the 16 inputs are sampled, using the internal or an external clock pulse to synchronise the sampling. These levels are stored, and in the usual operating mode, the triggering will cause the stored data to be centred around the triggering time.

For example, there may be 2K of data both before and after the trigger event. Triggering is an ANDed action so that you can set for some combination of signals that is unique, such as when the microprocessor writes a specified word to a chip during an interrupt.

Note that a display such as can be achieved using a logic analyser can also be obtained from a computer simulator. Simulator software allows the user to notify the chips and connections that are used in a digital circuit, and the program will then provide simulated waveforms. The value of this system is that any unwanted pulses (called 'glitches') can be detected, even to the extent of a 1ns pulse that in real life would occur once in 14 days, in the proposed circuit before it has been constructed, and the simulator can also be used to find such weaknesses in an existing circuit. One well-known simulator is PULSAR, from Number One Systems Ltd. of St. Ives; this has the advantage that it can be integrated with circuit diagram and PCB layout software.

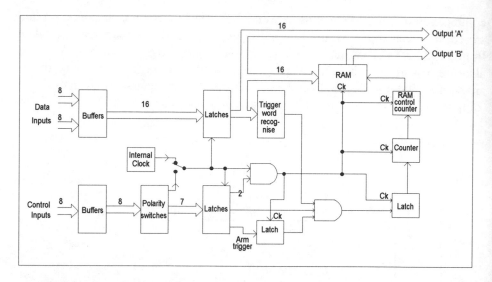

Figure 8.6 **A simplified block diagram for the data acquisition portion of the Thurlby LA-160 analyser, by kind permission of Thurlby Electronics**

Slow clocking diagnosis

For ordinary logic circuits, using TTL or CMOS chips, one very useful diagnostic method relies on slow clocking. The state of buses and other logic lines can be examined using LEDs, and with a one second, or slower, clock rate, the sequence of signals on the lines can be examined from any starting stage. This approach is not usually available for microprocessor circuits, but it can be provided by using an emulator, see below.

A few microprocessors of CMOS construction, like the Intel CHMOS 80C86, can be operated at very slow clock rates, down to dc. This is a very useful feature, though it does not necessarily help unless you can get the buses into the state at which the problem reveals itself. As in any other branch of servicing, your work is made very much easier if you have some idea of where the fault may lie. For example, if the user states that short programs run but long ones crash, this is a good pointer to something wrong in the higher-order address lines, such as an open-circuit contact on an IC holder for a memory chip. Slow clocking is not necessarily helpful for such problems, because a large number of clock cycles may be needed to reach the problem address by hardware methods.

Diagnostic programs

For many aspects of fault-finding, the use of a diagnostic program is very helpful. Such programs are usually available for computer servicing, and will help to pinpoint the area of the problem. Diagnostic programs are not always available for machine control systems, because so many systems are custom designed. If it's likely that one particular system design will turn up several times, then a simple diagnostic program should be written, enlisting a software specialist if necessary.

For small computers, the use of a good diagnostic program may be all that is normally needed to locate a fault. This is particularly true when the machine is one that has a reasonably long service record, with well-documented problems, and their solutions, available from a large sample of machines. Servicing the BBC Micro, for example, is made much easier by the service history which has been built up by local education authorities on this (now old) machine. Similar experience is available for the PC type of the machine and the Apple Mac, and for a few others.

In-circuit emulator (ICE)

For machine control circuits, easy availability of either data or spares cannot be relied on, and servicing may have to be undertaken with little more than a circuit diagram and the data sheets for the chips that are used. The compensation here is that the chips are more likely to be standard types, with no custom-built specials. The main problem in this respect is that servicing of microprocessor circuits cannot ever be a purely hardware operation. Every action of a microprocessor system is software controlled, and in the course of fault diagnosis a program must be running. For a computer system, this is easily arranged, but for a machine control system it is by no means simple.

A machine control system, for example, may have to be serviced in situ, simply because it would be too difficult to provide simulated inputs and outputs. In such circumstances, dummy loads may have to be provided for some outputs to avoid unwanted mechanical actions. Against this should be laid the point that inputs and outputs are more easily detectable, and likely to be present for longer times, making diagnosis rather easier unless the system is very complicated. A logic diagram, showing the conditions that have to be fulfilled for each action, is very useful in this type of work.

An in-circuit emulator is a particularly useful service tool for this type of work, and a block diagram is illustrated in Figure 8.7. The plug from the emulator replaces the microprocessor in the controller, and the emulator itself is a

133

computer circuit that uses a microprocessor (not necessarily of the same type), together with ROM and RAM

The software of the emulator provides signals from the output pins that are identical to those from the corresponding pins of a real microprocessor. Because these signals are now under software control it is possible to run at different speeds, examine register contents, set breakpoints, suspend some actions, isolate some inputs or outputs and so on. These actions are not possible when a real microprocessor is being used, and the emulator can also be used very much like an analyser, displaying traces of activity on selected lines for several (typically 40) clock cycles before a breakpoint.

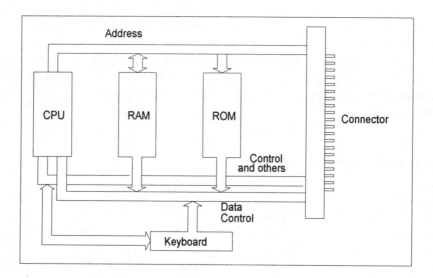

Figure 8.7   Block diagram for an in-circuit emulator

Using an emulator, the controller circuit and the devices connected to it can be checked with real signals. Some emulators allow sections of the controller ROM to be held in RAM, so that the controller software can be changed and tested. The codes in the RAM can then be burned into a PROM to make a new controller PROM.

An emulator must either be designed specifically for the equipment with which it will be used, or must be totally programmable. Some emulators are designed to be used to simulate ROM or PROM, only; others are available for microprocessors.

Note: Do not confuse this type of emulator with the computer simulator type of program that will simulate the working of a circuit before the circuit is constructed.

Signature analyser

A *signature* in this sense means a hexadecimal word, and the signature analyser is a method of finding an error in any logic circuit, including ROM, EPROM or CMOS RAM. The action is that a hexadecimal word is produced from each output when a set of inputs is applied to all the possible inputs of the circuit.

The basic diagram of the essential portion of a signature analyser is illustrated in Figure 8.8. This consists of a serial register with feedback taken from four of its parallel outputs to an XOR input. The feedback is applied so that any typical set of inputs will produce a set of bits in the register which is almost certainly unique — the principle, based on information theory, is that a register is working most efficiently when all states are equally likely (an n-bit shift register produces $2^n - 1$ different output patterns).

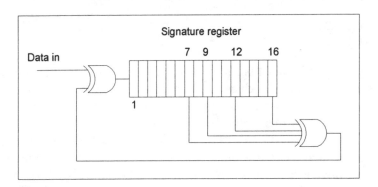

Figure 8.8   The register with feedback that is the basis of a signature analyser

Without feedback, a 16-bit register would produce a hex word that would simply reflect the last 16 bits fed into it. For example, if you fed in 16 1s, the register word would be FFFFH, and this would not alter if you continued to feed in 1s. By contrast, with the feedback connections illustrated, the action of feeding in 1s will produce a different register word for each 1 bit fed in, and there will be no repetition until 65536 bits have been fed in. This is illustrated in table 8.1 which shows 45 words that are generated when a stream of 1s is fed in — the first two words are omitted because they are simply 0001H and 0003H. The effect of the feedback starts to become evident after seven 1s have been fed in.

135

Table 8.1   Register (hex) words generated by a stream of level 1 bits

No. of 1s	Word	No. of 1s	Word	No. of 1s	Word
3	0007	18	F9CC	33	72A2
4	000F	19	F399	34	E545
5	001F	20	E733	35	CA8A
6	003F	21	CF67	36	9515
7	007F	22	9CCE	37	2A2B
8	00FE	23	399C	38	5456
9	01FC	24	7339	39	A8AC
10	03F9	25	E672	40	5159
11	07F3	26	CCE5	41	A2B3
12	0FE7	27	99CA	42	4566
13	1FCE	28	3395	43	8ACD
14	3F9C	29	672A	44	159A
15	7F39	30	CE54	45	2C34
16	FE73	31	9CA8	46	5669
17	FCE6	32	3951	47	ACD2

This register arrangement is also used to produce 'random' numbers, and when larger registers are used the chance of recurring numbers becomes less. In the 16-bit example, the words that are generated from 65,536 or fewer entries are therefore unique, and this allows us to trigger the start and stop of a signature analyser from the highest order of address line (A15) of an 8-bit microprocessor system, or from the A15 line of a system that uses more than 16 address lines.

Figure 8.9 illustrates the block diagram of a signature analyser with its inputs of clock, start, stop and data. The data input to the analyser will be from any line that produces signals, so that address or data lines can be used. Contact can be made by a probe or a clip. The register is also gated so that it operates for a fixed 'window', a defined number of clock cycles. In use, the system under test is 'exercised' meaning that inputs are applied — preferably so that all possible signal values are used at the inputs. The register of the signature analyser will fill with bits in the gated time window, and when the gates have closed, the four hex numbers displayed by the register form a word that should be unique to that combination of inputs and the point that is being tested.

The signature analyser must be used along with software that ensures that the inputs will be consistent, and a system that is known to be perfect is used to provide the signature words. If a system on test displays a different signature this will point to a fault, and the position of the fault can be found by repeating a signature check at different points along a signal path. One merit of signature analysis is that the same type of fault at a different point in a circuit will usually produce a different signature word.

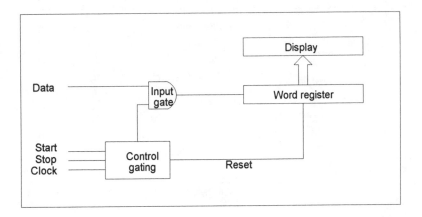

Figure 8.9 **The block diagram of a signature analyser**

Testing of combinational circuits is straightforward, but sequential circuits can be tested only if all feedback connections are broken.

Microprocessor circuits are usually signature tested by inserting an adapter between the microprocessor and its socket. The adapter is designed to allow pins to be open circuited or connected to either logic level so as to prevent interrupts or to isolate data lines from memory, and also to set a fixed data input such as the no-operation (NOP) code which will allow the clock to cycle the address lines continually. The probe can then be used at each point where a signature is known.

Figure 8.10 illustrates the modifications that have to be made to a Z80 circuit in order to carry out signature analysis. The interrupt line is broken so that the interrupt pin can be taken high, and the data bus is also broken so that the NOP command is always on the microprocessor data pins. These changes can be incorporated into an adapter that fits between the microprocessor chip and its normal socket.

A more restricted meaning of signature is applied to the ROM of a computer. The signature can be found in a variety of ways, such as by adding up all the stored byte numbers and taking the remainder after dividing by some factor — whatever method is used will have been devised so that the resulting word is unique and will be found only if the ROM contents are uncorrupted. The word that is obtained in this way will also be stored in the ROM, so that if the two do not match there must be a ROM error. This system can be used also on RAM when the RAM is filled with specified bytes.

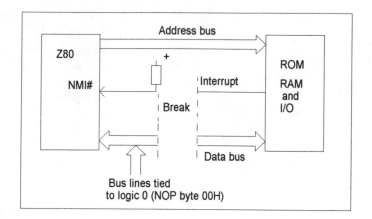

Figure 8.10   Signature analysis modifications to a Z80 circuit

Exercise 8.1

Construct a 16-bit serial register (using two 8-bit chips) and add the feedback connections from outputs 7, 9, 12, and 16 into an XOR (or an adder with carry ignored) to as to produce the format of the signature analyser register. Find the signature words for different numbers of input pulses.

Other digital instruments

Digital voltmeter

ICs exist to provide for measurement of all the common quantities such as DC voltage, signal peak and RMS amplitude, frequency, etc. Most of the modern ICs of this type provide for digital representation, so that the outputs are in a form suitable for feeding to digital displays. The most common IC of this type is the digital DC voltmeter IC, and a brief account of its action provides some idea of the operating principles of many instrumentation ICs.

Referring to Figure 8.11, the voltmeter contains a precision oscillator that provides a master pulse frequency. The pulses from this oscillator are controlled by a gate circuit, and can be connected to a counter. At the same time, the pulses are passed to an integrator circuit that will provide a steadily rising voltage from the pulses.

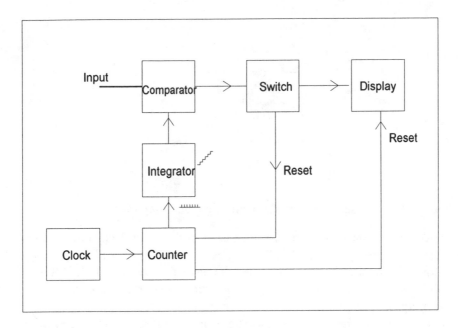

Figure 8.11   The principles of a digital voltmeter

When this voltage matches the input voltage exactly, the gate circuit is closed, and the count on the display represents the voltage level. For example, if the clock frequency were 1kHz, then 1,000 pulses could be used to represent 1V and the resolution of the meter would be 1 part in 1,000, though it would take one second to read one volt. The ICs that are obtainable for digital voltmeters employ much faster clock rates, and repeat the measuring action several times per second, so that changing voltages can be measured. Complete meter modules can now be bought in IC form.

Exercise 8.2

Construct a digital voltmeter from logic chips, using a 10kHz clock and making readings up to 10V. Use 100 pulses to represent 1V, and refresh at 0.25 second intervals.

Frequency meter

A frequency meter makes use of a high-stability oscillator to provide clock pulses, normally crystal controlled. In its simplest form, the block diagram is as shown in Figure 8.12, with the unknown frequency used to open a gate for the clock pulses. The number of clock pulses passing through the gate in the time of

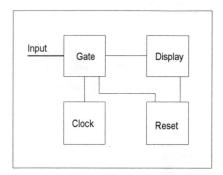

Figure 8.12 Block diagram for a simple frequency meter

one cycle of the input will provide a measure of the input frequency as a fraction of the clock rate. For example, if the clock rate is 10MHz and 25 pulses of the unknown frequency pass the gate, then the input frequency is 10/25MHz, which is 400kHz. A counter circuit can find the number and display it in terms of the clock frequency to give a direct reading of the input frequency.

In this simple form, the frequency meter cannot cope with input frequencies that are greater than the clock rate, nor with input frequencies that would be irregular sub-multiples. For example, it cannot cope with 3.7 gated pulses in the time of a clock cycle. Both of these problems can be solved by more advanced designs. The problem of high frequencies can be tackled by using a switch for the master frequencies that allows harmonics for the crystal oscillator to be used. The problem of difficult multiples can be solved by counting both the master clock pulses and the input pulses, and operating the gate only when an input pulse and a clock pulse coincide. The frequency can then be found using the ratio of the number of clock pulses to the number of input pulses.

Suppose, for example, that the gate opened for 7 input pulses and in this time passed 24 clock pulses at 10MHz. The unknown frequency is then $10 \times 7/24$MHz which is 2.9166MHz. Frequency meters can be as precise as their master clock, so that the crystal control of the master oscillator determines the precision of measurements.

Counter/timer

A counter/timer is used for the dual roles of counting pulses and measuring the time between pulses, and Figure 8.13 shows the block diagram, omitting reset and synchronising arrangements.

The counter portion is a binary counter of as many stages as will be required for the maximum count value. The input to the counter is switched, and in the

140

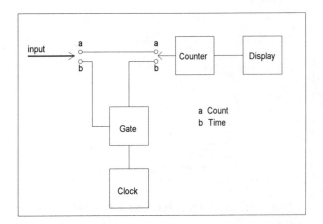

Figure 8.13   A simplified block diagram for a simple counter/timer

counting position, the input pulses operate the counter directly. When the switch is changed over to the timing position, the input pulses are used to gate clock pulses, and it is the clock pulses that are gated. In a practical circuit, the display would be changed over by the same switch so as to read time rather than count.

Exercise 8.3

Construct a clock oscillator circuit, using both the CR and crystal circuits illustrated in Figure 8.14. Measure the output frequency in each case. The circuits show 7414 inverters being used — these are particularly suitable because they have Schmitt trigger characteristics that prevents unwanted high-frequency oscillations from appearing.

Exercise 8.4

Construct a simple frequency meter, using the block diagram illustrated in Figure 8.13. Check the action, remembering that the readings will not be precise because only whole cycles are counted and no synchronisation is used.

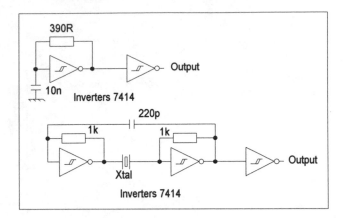

Figure 8.14 **Clock oscillator circuits. Use a crystal in the 5MHz to 16MHz range**

Test Questions

1. What is meant by (a) data domain analysis, (b) time domain analysis, referring to a logic analyser. What display would be used for each?
2. The following set of ten 8-bit binary signals have been stored in a logic analyser. Write the hexadecimal equivalent as it would appear on a word display, and sketch out the corresponding time domain signals.
    11010010   11001011   11000011   11001000   11000000
    10001101   10000001   10001010   10000010   10101011
3. Describe how you would use (a) a logic probe, (b) a logic pulser.
4. In a digital voltmeter, why is a slow clock used to reset the action at intervals?
5. Why is a simple oscilloscope inadequate for trouble-shooting logic systems?
6. How would a commercial frequency meter deal with a frequency which was an irregular fraction (such as 0.4632) of the clock rate?
7. In a clock circuit, one chip is used to provide an output and is not part of the oscillating circuit. What is this part termed and why is it used?
8. Why are Schmitt trigger stages often used in input stages?
9. A ROM signature is quoted as 7F4BH. What does this mean, and how could you check it?
10. Why is an in-circuit emulator so often used for diagnostic work on controller circuits for machinery?

9   Peripherals and input-output devices and circuits

Syllabus sections: D05–5.1, D05–5.2, I1–1.3.6, I1–1.3.7, I1–1.4.1, M04–4.1, M04–4.2.

Before the computer can usefully take part in the control of a system, it must be equipped with suitable interfaces to allow the parameters of the disparate external devices to be matched to those of the computer. The incompatibilities that arise include: different voltage, current or impedance levels; the computer and peripherals may work at widely different data rates; and the code formats of the computer and the external device may be very different.

Bus drivers/transceivers

In general, the interface circuits are specific to a particular type of operation. However the difficulties can be minimised by employing special programmable interface ICs referred to as peripheral interface adapters (PIA) or parallel input/output devices (PIO). Tri-state logic switches shown in Figure 9.1(a & b) are used to ensure that no more than two elements of the system can use the bus at any one time. Furthermore, Figure 9.1(c) shows how transceiver devices are used as buffers to transfer data across the bus system under the control of a single bit that can be set to 0 or 1.

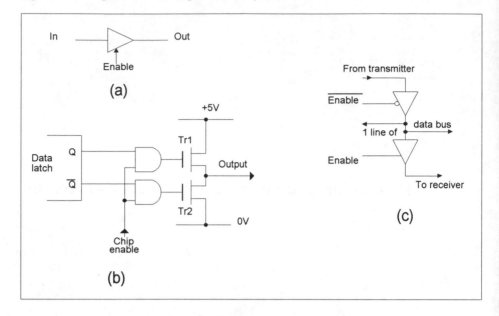

Figure 9.1   Tri-state logic and bus transceiver control: (a) circuit symbol, (b) implementation of device, (c) application to bus control

Exercise 9.1

Use either a logic trainer or suitable microprocessor control board to study the behaviour of tri-state logic gates.

Clock control circuits

All operations and data transfers have to be accurately time controlled and Figure 9.2 shows a number of oscillator circuits that employ digital ICs as the active component. The operation of these circuits is often based on the propagation delay through an invertor or other gate devices.

In Figure 9.2(a), the inverting Schmitt trigger circuit behaves as an astable multivibrator to produce a near square wave output with a frequency that depends on the values of R_1 and C_1.

With a value of 390Ω for R_1, the frequency will range from about 20Hz to 2MHz as c_1 is varied from 100pF to 1nF. The circuit of Figure 9.2(b) depicts an ideal inverter driving an RC network. The chip output impedance which is largely responsible for the propagation delay is represented by R_1C_1 and the delay is dependent on the rate of charge and discharge of C_1 through R_1. This basic

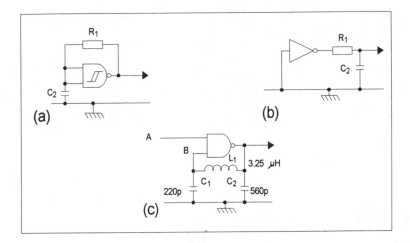

Figure 9.2   Some typical clock oscillator circuits

concept is used to produce a ring oscillator by series coupling an odd number of inverters and employing feedback over the whole loop. For a series of n inverters each with a propagation delay (T_p), it will take nT_p seconds for the output device to change from the High to Low state. Similarly it will take a further nT_p seconds to change from Low to High.

The cycle time is thus $2nT_p$ seconds and from this, the frequency of oscillation is given by $1/2nT_p$ Hz. The frequency can be reduced by increasing the number of inverters in the loop or by introducing a further delay in the form of an RC network as shown in Figure 9.2(b). If the gate inputs A and B of Figure 9.2(c) are connected together, the circuit oscillates at a frequency controlled by the series combination of C_1 and C_2, in parallel with L_1. The component values shown, produce an oscillation frequency of about 7MHz. By coupling gate input A to a serial data stream instead, the oscillations become locked to the data rate of about 7MHz. Often a spare inverter stage is included as a buffer amplifier with these circuits to reduce any loading effects. A more frequency stable and common circuit using this concept, was shown in Figure 8.14.

Optical coupling

Many peripheral devices have to operate some distance from the central computer and the data may have to be transmitted through an electrically noisy environment. In these cases it is possible to ensure the data integrity by using an

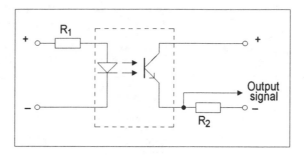

Figure 9.3   Opto-coupler/isolator

optical fibre communications link. Figure 9.3 shows the principle of operation of an opto-coupler/isolator device. Since a light beam is unaffected by electromagnetic interference and has no electrical impedance, it will not be influenced by radiation or introduce a mismatch situation.

The opto-coupler/isolator consists of a light generator and a photo-detector housed in a closed, light-proof housing. When an input signal is applied to the LED, the resulting light output will modulate the current through the photo-detecting transistor to provide the isolated output signal. Depending on the nature of the load to be driven, the circuit can employ a photo-detector that consists of a Darlington pair for greater gain, a thyristor to handle high current levels, or even a simple diode. For communications over a greater distance, the light-proof housing can be replaced by an optical fibre with the appropriate couplings at either end.

Voltage level shifting

It is often necessary to equalise the voltage levels between coupled stages in cascade. Examples include the coupling of TTL to CMOS stages and the 5V to 3.3V level shift needed for low voltage digital processing devices. Such a coupling stage may also need to employ the unity gain and impedance matching properties of an emitter follower as a buffer.

Referring to Figure 9.4(a), if the output is taken directly from the emitter, V_o will be 0.7 volts less than V_i. This value can be increased by adding an extra resistor R_2 as shown. The voltage shift is then lifted by the volts drop across R_1. However, this has the disadvantage of introducing a potential divider at the output which will attenuate any ac signal component. This can be overcome by replacing R_2 with a second transistor biased as a constant current device to increase V_o by $I_o x R_1$ and without the signal attenuation.

146

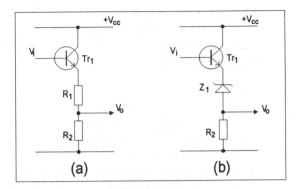

Figure 9.4   Voltage level shifting circuits

A further modification involves adding a series zener diode as shown in Figure 9.4(b). Provided that V_{cc} is high enough, then $V_i - V_o = 0.7 + V_z$ volts. Alternatively, the zener diode can be replaced by a string of forward biased diodes, each with a threshold level of 0.7V. Four such diodes then behave as a 2.8V zener diode.

Driving inductive loads

The maximum output current of a TTL gate is 10mA and with 5 V drive, this produces a maximum power of 50mW. Since most electromagnetic devices require a drive considerably greater than this, then an interface will be needed between the TTL part of the circuit and any useful inductive load. Figure 9.5 shows how a buffer amplifier can be added to meet these requirements.

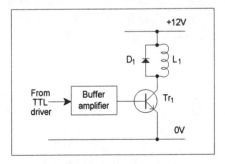

Figure 9.5   Driving an inductive load

The load L_1 may be a relay, a solenoid operated valve, an electric motor or other similar device. When such a circuit is switched to the *off* state, a large back emf is developed across the load which can momentarily forward bias the collector-base junction and burn out the transistor. This is prevented by the addition of D_1, the catching or flywheel diode, which becomes forward biased at switch-off and damps the back-emf pulse.

Driving the stepper motor

The construction, operating principles and programming of these motors was covered in Volume 2, Part 3 (Control Systems) of this series, to which the reader is referred for useful revision. Since each stepper motor contains multiple stator windings, each will need an interface similar to that of Figure 9.5.

By the nature of its construction, the stepper motor has an impedance that is highly inductive with a low effective series resistance. The L/R time constant is therefore fairly high so that input pulses become distorted. This has the effect of reducing the motor torque, particularly at the higher pulse speeds. A series resistor is therefore commonly added to each winding to lower the time constant and improve the motor torque output.

Exercise 9.2

Use an oscilloscope to study the drive waveforms associated with a stepper motor when connected and programmed as indicated in Chapter 5 of Volume 2, Part 3 of this series.

Driving the servo motor

As explained in Volume 2, Part 3, the servo motor is designed to aid mechanical positioning and is therefore usually included within a feedback loop. The motor is usually operated as a shunt wound unit with the armature driven from a constant current source and the field from an amplified version of the system error signal. Again, this motor represents a highly inductive load for any drive system and should be treated as described above.

Shaft encoders or digital resolvers

As explained in Volume 2, Part 3, the shaft encoder functions on the opto-coupler principle to directly generate a digital output signal due to rotation. The device is mounted on a rotating shaft and consists of a disc that carries etched gratings which interrupt the light beam. There are two types of encoder in use, the incremental and absolute devices. In the former, the disc carries a series of peripheral slits that generate a continuous bit stream due to rotation. The absolute shaft encoder uses a disc engraved with a number of concentric tracks, each with its own opto-coupler to produce a parallel bit output pattern. A practical incremental shaft encoder carries three tracks, one that generates a reference pulse once per revolution, and two that produce typically 512 pulses per revolution which are staggered by a quarter period as indicated by Figure 9.6.

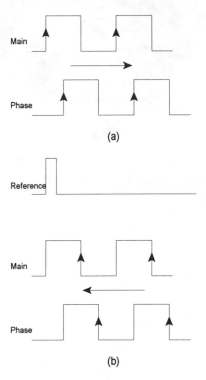

Figure 9.6   Incremental shaft encoder waveforms: (a) forward, (b) reverse

This version requires the addition of a counter circuit to determine both the angle and number of revolutions. By using suitable logic, the timings of the positive going edges which change with direction of rotation, can be compared to detect the direction of rotation. See Figure 9.6(a & b).

Figure 9.7, overleaf, shows a simple 4-bit absolute shaft encoder disc which requires four accurately positioned opto-couplers to generate its output signal. Because of possible positioning errors and the fact that rotation can stop at the junction of two consecutive segments, errors can arise in the output count. Therefore, all absolute encoder discs are produced in a Gray code which is cyclic and only changes one bit at a time. Any errors are then limited to 1 bit only.

Table 2.2 (Chapter 2) has been used to code the disc in Figure 9.7.

Figure 9.8 shows how the Gray code can be converted into binary using either a ROM look-up table where the Gray code bits are used as an address, or by the direct use of X-OR logic.

Figure 9.7   4-bit absolute shaft encoder

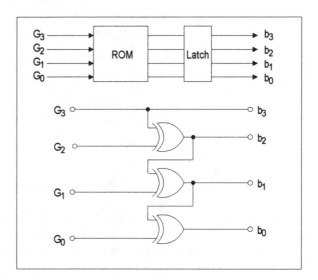

Figure 9.8   Gray to binary coder converter

Switch contact bounce

All switch mechanisms with contacts mounted on springy levers have a self resonant frequency. Therefore when the switch is operated with a sudden impulse, the resulting natural vibration sets up contact bounce, which can last for a few milliseconds before settling down to an equilibrium state. The effect adds a number of false pulses at the moment of contact make which can cause a digital system to malfunction.

The effect can be eliminated by using an RS flip-flop as shown in Volume 2, Part 1, or with a Schmitt trigger circuit as shown in Figure 9.9. Here the spurious signals introduced by contact bounce fall outside of the Schmitt switching threshold and so are clipped off. The CR time constant is so chosen that the circuit does not respond to the transitions introduced by the contact bounce. For circuits with normally closed contacts, the positions of C and R are reversed.

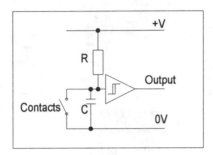

Figure 9.9   Debouncing switch contacts

The keyboard

The complex keyboard consists of a matrix of row and column lines with a keyswitch at each intersection, plus a keyboard controller IC. Each key press generates a unique code that is used to address a ROM memory. This is programmed to hold all the necessary output codes, such as ASCII etc. The two common matrix arrangements in general use, are either 8×11 (88 key board) or 8×13 (102/104 keyboard) and each requires a dedicated controller IC which is commonly a 40-pin chip.

Figure 9.10 shows the basic principle of operation with the ROM being addressed from an m stage matrix (m = 11 or 13) and an 8-stage ring counter. Whilst no key is depressed, logic 1's continually circulate around each counter, so that each ROM location is being addressed in a cyclical manner. The absence of a strobe output at this stage signifies that the data output is *not valid*. Each key contact in the matrix then corresponds with a unique location in the ROM. The count rate is controlled by the internal clock that can be externally programmed by C_2R_2, to run at 50KHz or 100KHz.

When a key is pressed, a path is created from one output of the 8-stage counter to one input of the m-bit comparator. After a number of clock pulses, the m-stage counter will produce a matching bit pattern as the second input to the comparator, indicating that the addressed ROM location corresponds with the key that has been pressed. This match condition stops the clock and generates a strobe output

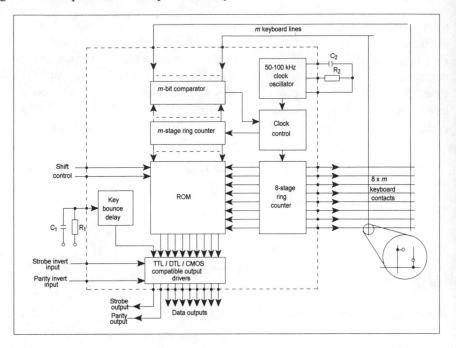

Figure 9.10   Multiplexed keyboard

via the delay network. The 9-bit contents (8 ASCII bits plus parity) of the ROM address are now loaded into the output latch and the strobe signal implies that the data is valid. The data outputs remain stable until the key is released. The single delay network externally controlled by C_1R_1, is designed to suppress contact bounce on all keys.

By suitable use and combination of the shift and control keys, and provided that enough ROM space is available, up to four different character sets can be generated. Many keyboard controllers have the facility for user selection of odd or even parity and change of output polarity.

Two other useful features include N-key lock-out and 2-key rollover. In the former case, if two or more keys are pressed during the same clock cycle, only the first is valid. The other keys being locked out until the first one is released. By comparison, 2-key rollover allows two keys to be validly operated in quick succession.

Seven-segment displays

The basic constructional and operational features of these display devices is included in Volume 2, Part 1, of this series to which the reader is referred for such information. Although described as seven-segment display devices, these

152

commonly have eight illuminating segments due to the need to include a decimal point. This may appear either in front or behind the main digit.

Light emitting diode types (LEDs)

To minimise the number of contact pins needed, each diode anode or cathode is returned to a common pin. Devices may therefore be described as common anode or common cathode types. In the case of common anode devices, this pin is returned to the positive supply and the cathodes individually driven to logic 0 to illuminate. Conversely, common cathode device anodes are driven to logic 1 for display purposes. The circuit for each segment contains a current limiting resistor to ensure that the diode current does not exceed about 25 to 30mA.

In order to save power, particularly with battery driven equipment, a multiplex or strobing technique is employed. In this case, a series of display devices may be driven On and Off in sequence at a fast enough rate to give the appearance of continuous illumination. This concept is shown in Figure 9.11 together with the multiplex and illuminating waveforms for a pair of common cathode devices.

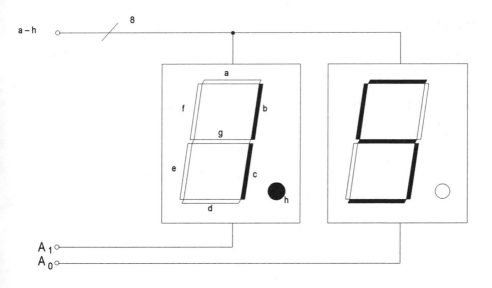

Figure 9.11   Two digits of a multiplexed common-cathode display

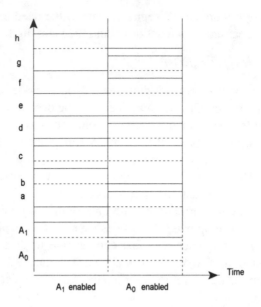

Figure 9.12   The time-related waveforms for the display of Figure 9.11

In order to display either 10 decimal or 16 hexadecimal characters, it is necessary to use a 4-bit code sequence, either BCD or its extended hexadecimal version. Because there is no obvious relationship between the input codes and the required display form, it is necessary to use a decoder circuit as shown in Figure 9.13.

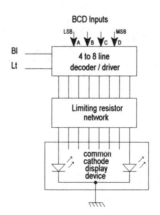

Figure 9.13   BCD to seven-segment display decoder/driver

Although these devices are usually termed 4- to 7-line decoders, they also provide for the illumination of the decimal point. In the circuit shown, whenever an output line of the decoder is driven high, the corresponding segment is illuminated. The control line marked Bl is used for blanking purposes to allow the unwanted leading zeros to be suppressed. The line marked Lt provides a light test facility. When this line is taken low, all segments are illuminated.

Figure 9.14 shows an example of the way in which a multiplexed seven-segment display can be driven from a microprocessor via a dual port interface adapter under software control. The 8 bits output from port A are used to drive the display segments and the decimal point. When a port B0 to B3 line is driven to logic 1, the corresponding display is illuminated. For multiplexing, the four least significant cells of port B data register must be loaded cyclically with the sequence 0001, 0010, 0100, 1000, 0001 etc.

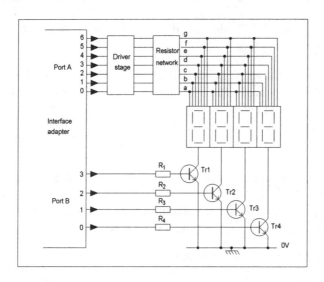

Figure 9.14   Multiplexed seven-segment display

Liquid crystal displays (LCDs)

The principles and construction details of LCD panels are explained in Volume 2, Part 2, of this series. However, seven-segment type display devices are also available and these are popular because of their low power consumption. Each segment bar is formed by the deposition of a transparent layer of indium tin oxide on the inner surface of a glass plate. The space between this and a similarly coated backplate glass is filled with the liquid crystal material.

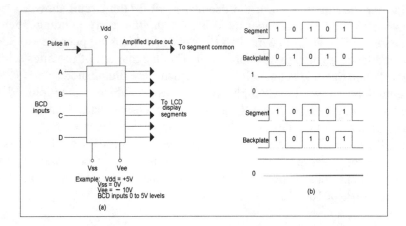

Figure 9.15 (a) LCD device driver, (b) PD between segments and backplate

Figure 9.15(a) shows a typical BCD to seven-segment LCD device driver, together with the typical supply potentials. Conductor patterns to each segment are also fabricated in indium tin oxide. When an electric field is applied between a segment and the backplate, this causes a change in the molecular structure of the crystal and produces an opaque (black) display. Without such a PD a segment is transparent.

Because any dc applied across the crystal material produces a progressive degradation of the display quality, LCDs are driven by squarewave ac signals. Figure 9.15(b) shows how the potential difference needed to drive the electric field is derived between a segment and the backplate.

Modems (MOdulator-DEModulators)

The plain old telephone system (POTS) forms a popular medium for communications between computer terminals and in spite of its restricted audio bandwidth (300Hz to 3.4kHz), can achieve a surprisingly high data rate. In the past, both FSK and PSK have been used extensively for low bit rate operations, but today, some smart modems with elegant modulation schemes, can operate with bit rates as high as 29.2kbit/s. This is achieved by combining the characteristics of ASK and PSK, whereby each transmitted symbol can represent a number of bits.

The method is referred to as *quadrature amplitude modulation* (QAM), where each of 4, 8 or 16 unique phases can be allocated up to four different amplitudes. 64QAM is in common use and this allows each instantaneous symbol to represent

156

6 bits of information and very significantly reduce the transmission bandwidth while still maintaining a high bit rate. 256QAM has been introduced for special applications and this provides 1 byte of data per transmitted symbol.

These smart modems always start up at the highest bit rate available and by using a line probing signal to test the signal to noise ratio, finally adopt the highest speed acceptable to both communicating modems. With those modems that can implement data compression algorithms, effective bit rates as high as 150 kbit/s have been achieved. Because the communicating terminals and the POTS use different power levels, each modem is equipped with a transformer barrier unit to provide electrical isolation. Furthermore, because the transmission occurs over a single pair of wires and each terminal requires two lines each for transmit and receive, a second transformer is used as a 2-line to 4-line coupling device. This transformer, known a *hybrid* also acts as an echo-cancelling device to minimise the cross-talk between transmit and receive circuits.

Exercise 9.3

Use a modem to transmit and receive data from another computer. Include a break-out box at one end of the system and study the handshake signals. Study the data signal formats for different data rates.

Disk memories

Floppy disks

The basic element of this range of devices, is a thin plastic (mylar) disc that is coated with a magnetic material and permanently enclosed in a low friction envelope. Disk casings of 3″, 3.5″, 5.25″ and 8″ width are available. When placed in its drive, the disk is clamped to a spindle and rotated (typically) at 360 rpm. When the drive is selected, a read/write head permanently contacts the disk surface through a slot cut in the casing.

The magnetic surface is divided into a number of circular tracks and then further sub-divided into sectors in the manner shown in Figure 9.16. The number of tracks and sectors depends upon the recording standard in use. An index hole in the disc which is sensed by an opto-electronic detector, is used to indicate the start of each track.

Sectoring is controlled in one of two ways. On the larger disks *hard sectoring* is common, with index holes used to indicate the start of each sector. The smaller discs use *soft sectoring* where only the start of the first sector is indicated, the others being determined by division within the disk handling software.

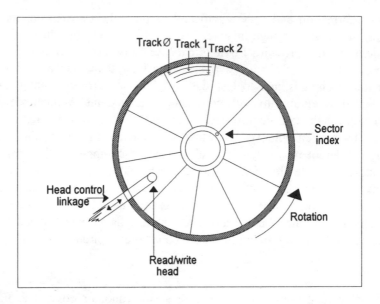

Figure 9.16   Sectoring of a floppy disk

The data is recorded on to the disk surface, by a bit stream that is derived from a series of clock and data pulses. This uses a format that is based on a frequency shift keying (FSK) waveform that drives the magnetic material into saturation. The so-called FM method is based on a primary coding format generated according to the following algorithm:

write a clock pulse at the start of each bit cell;

write a second bit pulse at the centre of the cell if and *only if*, the data bit is logic 1.

The preferred second method which provides a greater data packing density is commonly referred to as f-2f or MFM (modified or modulated FM). The rules for this algorithm are as follows:

If and only if the data bit is logic 1, write a data pulse at the bit cell centre;

If the previous bit was logic 0 AND the current bit is also 0, write a clock pulse at the start of the bit cell, otherwise write nothing.

The data capacity for each disk depends upon the diameter, the data packing density that can be achieved by the read/write head construction, the coding format used and the disk to head speed. Typically a 3½" disk will use either 40 or 80 tracks per side and be organised into an odd number (often 9) of sectors per track and either 128, 256 or 512 bytes per sector. The highest of these figures will lead to a maximum unformatted data capacity of 1.44Mbits. The data transfer rates range from 125kbit/s to 250kbit/s for FM or MFM recording respectively.

Before a disk can be used to store data it must first be formatted. This operation segments the disk by laying down appropriate magnetic patterns to provide the necessary track and sector addresses and synchronism.

The other important parameters of the floppy disk system are related to time delays. These include:

Seek time — the time taken to position the head on the selected track.

Latency — the time taken for the selected data sector to reach the head once it has been positioned on track. Average latency is about 100ms.

Access time — the time taken to start reading or writing to a selected track.

Stepping time — the time delay for stepping between adjacent tracks, typically about 20 ms.

Total access time — the sum of latency and seek times.

Average access time — half the worst case total access time, typically about 200ms.

The floppy disk controller

The disk drive unit contains two motors, one a dc device used to run the disk at a constant speed, and the other a stepping motor to control the position of the read/write head. The dc motor is normally included in a servo system feedback loop in order to maintain the constant speed. Figure 9.17 shows the general arrangements and signals associated with this interface.

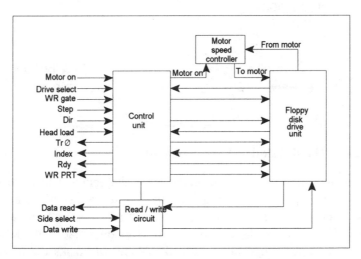

Figure 9.17   Floppy disk controller

Some of the signals which are summarised on the following page are two-state and act as switches whilst others are generally active low.

Motor on —- an active low signals the disk drive motor to run.

Drive sel — selects one of several connected drives.

WR gate —- a low/high signal selects write/read operation respectively.

Dir — direction. Low, head steps inwards; high, head steps outwards.

Step — head steps on positive edge of signal.

Head load — active low energises a solenoid so that the head contacts disk surface during read or write operations only.

TRØ — low when head is over track zero.

INDEX — active low when the index hole has been located.

RDY — active low to indicate a particular drive unit is ready.

WR PRT — write protect, active low, prevents over-writing of previous data.

Side select — selects side 1 or 2 on double-sided disk drive systems.

Data Read/Write — selects the direction of the data transfer.

Hard disks (Winchester technology)

Typically this technology is based on disks ranging in diameters from 1.8", 2.5", 3.5", 5.25", 10", through to 12", which are assembled into either single or multi-platter stacks as indicated in Figure 9.18.

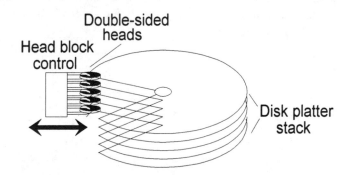

Figure 9.18   Hard-disk drive arrangement

The larger and earlier disks were made from aluminium whilst the later and smaller versions may be made from glass. The disk sets are clamped to a hub in a permanent stack and then enclosed in a sealed housing to form a dust free environment. Each disk or platter surface carries its own read/write head and the assembly is only interchangeable as a pack. Aluminium disks are nickel plated and highly polished to minimise surface imperfections. Each surface is then coated with a fine film of magnetic ferric oxide or cobalt to provide the memory element.

Each set of corresponding tracks on each surface is referred to as a *cylinder*. Thus with a multi-head and platter configuration, a particular track is selected by

160

a cylinder and a head number. Usually one surface, known as the *servo surface* is reserved for speed control of dc the drive motor and moving-coil type of position control of the heads. This type of mechanism is preferred because it is self-parking when power is switched off. Unlike the floppy disk, the heads of hard disk systems do not generally make physical contact with the disk surface, but fly a few microns above it. When the disk is at rest, the heads are usually parked on the inner cylinder surface. The glass substrate is chosen because it is harder and has a smoother finish than aluminium which allows a reduced flying height to achieve a higher data packing density.

In operation, the disk assembly rotates at a typical speed of 3,600 rpm to give an average access time to the data of about 10 to 15ms. A very large data capacity is available, depending chiefly upon disk size, this can range from about 100Mbits up to 9Gbits. Because of the high degree of mechanical precision necessary, the larger disk drives and packs are expensive, though prices have dropped dramatically in 1995. The cost per bit for storage is therefore relatively cheap. By comparison, the smaller Mini-Winchester drives are available at a much lower cost and still provide a data capacity approaching 1Gbyte.

Some of the more important hard disk parameters are as follows:

Rotational speed — 3,600, 4,500, 5,400, or 7,200 rpm.

Disk motor spin-up time — less than 1.5s.

Track to track seek time — 2ms.

Average seek time — 11ms.

Average access time — 10 to 15ms (often depends upon interface).

Data transfer rate — 1 to 2Mbit/s (again depends on interface).

The MFM coding used by earlier hard disk drives, has largely been replaced by run length limited (RLL) coding. Since this restricts the number of consecutive zeros allowed in the coding process, it leads to a greater data capacity. For example, 2,7 RLL and 3,9 RLL, coding allows zero run lengths between 2 and 7 or 3 and 9 before inserting a false 1. This leads to a data capacity improvement of 50% and 100% respectively.

Hard disks for small computers can be *partitioned* during formatting to allow a large capacity disk to behave as though it were several smaller units. This was done originally because older operating systems could not cope with large disk sizes, but is seldom necessary nowadays.

The data capacity and access times are dependent upon the type of interface in use. Using the popular IDE (integrated drive electronics) system which allows a maximum of four additional devices to be added, the typical figures are 540Mbit per drive and 11ms respectively. By comparison, SCSI (Small computer systems interface), provides for up to seven additional devices, all of which can be addressable disk drives, to provide a total capacity in the order of 9Gbit. SCSI interfaces are used for the more powerful machines with large hard-drive

161

capacity, but the EIDE system (extended IDE) provides for up to four hard drives.

Exercise 9.4

Use an oscilloscope to verify the waveforms and interface signals at the output of a floppy disk controller.

Optical disks

This most recent and developing example of disk technology, is characterised by a very large data capacity and high packing density. These devices are available in one of two forms. The permanent *Write Once, Read Many* or WORM disk is effectively a read-only memory using the format of an audio CD, and the more recent erasable and re-recordable types are at the time of writing much more expensive.

Disks consist of a glass or plastic platter that carries a very thin sensitive vacuum-deposited layer on one side, to act as the memory. This layer is then coated with a transparent layer for protection and two such disks can be glued back to back to form a double sided memory. The data is recorded by a laser beam in a continual spiral track of pits. To ensure synchronism, some disk systems include a pre-recorded sync track. The general principle is indicated in Figure 9.19.

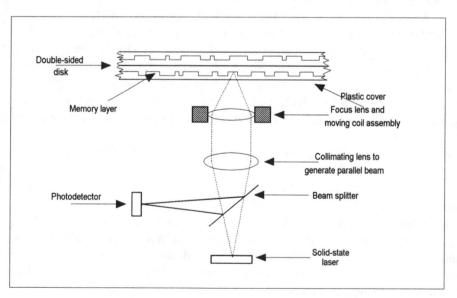

Figure 9.19   Principles of optical disk scanning

About 10mW of laser light energy is focused on to the sensitive surface to write the data. The exact effect depends upon the technology used. Reading is accomplished by using the same laser but at a much reduced power level of typically 0.5mW. The reflected light energy is split from the main laser beam and focused on to a photo-detector to regenerate the data. To ensure correct focus, a final lens in mounted in a moving coil assembly that is driven by part of the servo system. The drive signals for this and other positional control being obtained from the recovered data signal path.

Disks are available in several sizes ranging from 5″ to 12″ diameter, these being capable of storing more than 650Mbytes and 1Gbyte respectively. The 5″ version is described as a *CD-ROM* because the concept is closely related to that of the compact audio disk. Such a large capacity produces a longer average access time than other systems, but the typical 250ms is not too troublesome, as these memories tend to be used for archive purposes.

The standard speed CD-ROM drive provides an output bit rate of 125kbit/s, but double, quad and hex (*x*6) speed drives now produce bit rates of 300, 600 and 900 kbit/s respectively.

Optical disk recording technologies

Dye polymer technology. Certain organic dyes, including alloys of zinc and silver respond to light energy in a very convenient way. When a very thin layer is exposed to laser light, local heating causes microscopic pits or craters to be formed to provide a permanent memory. The reflectance from the surface now varies according to the pit pattern. The reflected light energy has a different and varying phase from the incident beam and so it can be extracted as the data signal. Other organic dyes change colour as the laser beam causes local spots to change between the crystalline and amorphous states. The colour changes again vary the reflectance. This latter technique produces erasable and reusable discs.

Heat sensitive technology. In this case, the sensitive layer is an alloy of Tellurium and Selenium with a light doping of Arsenic to give more accurate control of the melting point. During the *Write* mode, the laser beam burns microscopic holes in the sensitive layer and these are detected by the lower power beam during a *Read* operation. Erasure is achieved by using just enough laser energy to melt the layer without forming holes. The molten regions then cool very quickly to solidify into a stable amorphous state. Re-writing ensures final erasure by transforming the amorphous domains back into the crystalline state.

Magneto-optical technology (M-O). This technique is alternatively referred to as *optically assisted magnetic recording.* A thin layer of amorphous *gadolinium-iron-cobalt* is used for the memory and the whole surface is initially magnetically polarised in the same plane but at right angles to the surface. When heat from a laser beam is applied and simultaneous with a magnetic field from the

163

other side of the disk, this small region reverses polarity and becomes frozen in this state. When plane polarised light is reflected from a highly polished electromagnetic surface, the light becomes elliptically polarised. This is the Kerr Effect. A read operation is therefore performed using polarised laser light to detect the field changes. For erasure, the surface is again laser heated but with the applied magnetic field polarised as in the original sense.

Printers and plotters

These devices are designed to produce a hard copy output of the results of computer processing. The important parameters include the speed of output and the quality or resolution of the copy, properties which tend to be capable of being traded against each other. Whilst printers are chiefly intended to produce text outputs, plus a useful graphics capability, plotters are intended to produce much larger diagrams for engineering or architectural applications. Plotters which have been described in Chapter 3 of Volume 2, Part 3 of this series, produce line diagrams drawn by a pen driven simultaneously in two dimensions via two separate servo systems.

The design of printers intended for operation with personal and desk top computers varies considerably. They are capable of handling different types of paper as single sheets or continuous fan-fold, using friction drive or tractor feed and with a range of print quality and type fonts. The latter are generally capable of being changed whilst in operation, under software control. A range of printing technologies are employed for monochrome and colour purposes, using both plain and specially sensitised papers.

Printing is a relatively slow process and so most printers contain a buffer memory. This is periodically loaded with large blocks of data so that the processor can then continue with other tasks whilst the printer is emptying the buffer. The print quality ranges from Draft represented by a low dot resolution and suitable only for listing purposes, to a higher resolution that provides a near letter quality (NLQ) that compares favourably with a typewriter output. Whilst such printers operate on a character by character basis, many industrial printers are available that operate much faster on a line by line basis.

Most printers associated with personal or desk top computers employ a Centronics interface. This provides an 8-bit parallel port for relatively fast data transfers, together with a number of control or handshake lines. The control lines include:

Strobe — active low to transfer data.

Acknowledge — active low to indicate that data has been received and the printer is again in the ready state.

Busy — active high to indicate that printer is occupied, has detected an error or is off-line.

Select — active high to indicate that the printer has been enabled.

Reset — active low to indicate that the printer has been reset to its initial state and the print buffer has been cleared.

Error — active low to indicate that the printer is off-line, out of paper or has detected an error.

Select in — active low. Data entry is only possible under this condition.

Some printers are equipped with an optional interface to the RS-232D standard, but serial transfer tends to be a slower process.

Furthermore, printers can be sub-divided into two classes, the impact and non-impact types. Only the former are capable of producing carbon copies.

Daisy wheel printers

This is one of the impact class of devices. Whilst it is now virtually obsolescent, the daisy wheel printer, because of its typewriter characteristics, actually set the original letter quality print standard. A raised character is imprinted on the pad at the end of each of 96 spokes that are arranged in a daisy petal fashion as shown in Figure 9.20.

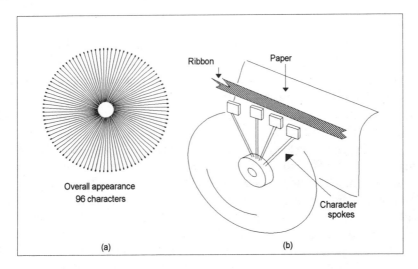

Figure 9.20 **Daisy wheel print head: (a) overall view, (b) detail of principles**

A stepper motor rotates the wheel to the desired character position and the spoke is then struck with a small electromagnetically driven hammer. The typeface then impacts the ribbon and paper against a platen to transfer ink from the ribbon to the paper in the typewriter fashion. Periodically the wheel must be rotated to its home position to ensure synchronism between the character data and the print wheel position.

Offset against the high print quality are the low speed, typically less than 20 characters per second. In addition, the printer does not have a graphics mode and changes of fonts or alphabets require a change of print wheel. This type of printer is virtually obsolete.

Ink/bubble jet printers

These two non-impact types utilise a reservoir or cartridge to contain the ink that is sprayed on to the paper. A major problem is to maintain the ink in a fluid state within the cartridge and then achieve quick drying once it is on the paper. The ink is directed on to the paper as a high velocity stream which is deflected by an electrostatic field, or broken up into bubbles through the action of an ultra-sonic transducer. As the cartridge traverses across the paper, ink from up to 16 vertically positioned very small apertures, discharge the ink on to the paper. This action provides a high resolution of print quality, equal to that of the laser printers, but at a lower cost. The drawbacks are the need for suitable paper, and the cost of renewing the ink-jet and ink-cartridge assembly at intervals.

Laser printers

A focused laser beam and a rotating mirror is used to scan, line by line, an image on the surface of a charged, photo-sensitive drum, typically producing 300 lines per inch. (See Figure 9.21). This mechanism is almost identical to those used for mechanical television in the period from 1888 to 1930.

The drum image which is in the form of a variation of electrostatic charges attracts and retains fine particles of black toner that are brushed over the drum. A sheet of plain paper is then passed over a very fine wire that carries a high voltage to impart a negative charge. This paper is then passed over the drum carrying the toner and this is now transferred to the negatively charged regions. The toner is finally fused to the paper by passing it over a heated roller, to provide a permanent record. The drum next passes over a second but positively charged wire (the corona wire) to neutralise its charge ready for the next printing. Although these printers are relatively costly, they produce print and graphics of typesetting quality, with resolutions of 300, 600 or even 1,200 dots/inch. The printing speed is fast, up to 15 (identical) A4 pages per minute (ppm) in monochrome and 3 ppm in colour.

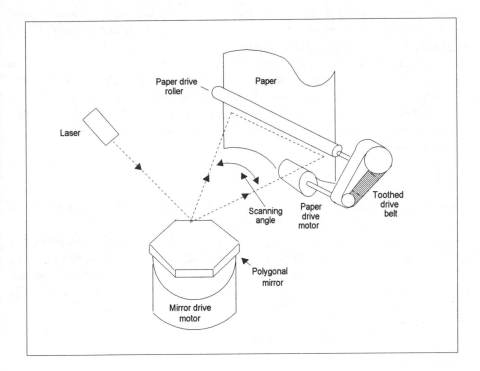

Figure 9.21   Laser printer drive mechanism

Dot matrix printers

These impact printers are probably the most popular for small computer systems. Although highly complex in construction and often employing an embedded microprocessor for control, they are relatively inexpensive due to the large scale of production. In some machines the print resolution can achieve letter quality and both fonts and alphabets are software programmable.

The print head carries 9 or 24 pins arranged in a vertical stack. These are fired by electromagnetic solenoids under data bit control to produce the character dot matrix which for a 9-pin machine is typically 5 x 9 in area. Since such a printer can generate $2^{45} = 3.52 \times 10^{13}$ unique character patterns, these machines have a very useful graphics capability.

The printer is driven by two stepper motors, one for paper feed and the other to drive the print head via a toothed flexible belt across the paper. In operation, the printing pins strike an inked ribbon against the paper to transfer ink dots. The pin return is achieved by the reaction of the paper against a platen, assisted by small springs. The paper feed motor drives the paper either by a friction roller or a

167

toothed sprocket wheel that forms a tractor drive. As the head carriage driving and timing belt rotates, it also drives a set of pinions to make the endless loop inked ribbon circulate through its cassette.

The printer mechanism carries a number of sensors, usually operating on the opto-coupler principle. These include, a *home sensor* to indicate the extreme left hand edge of the document, the *current head position sensor* to provide the cursor and an *end of paper sensor*. The print speed depends upon the selected printing mode, ranging from about 50 characters per minute in NLQ, to 200 characters per second in draft mode.

Exercise 9.5

To test a device handler routine for parallel data transfer. Assemble the system as shown in Figure 9.22 and carry out the following.

Machine code programs are provided for 6502, Z80 and 6800 microprocessor based systems. The aim of each program is to read a data byte input from Port B, invert the data and then output the byte through Port A. The examples contain references to both memory mapped and non-memory mapped I/O devices. In each case, the first step is to initialise the ports as input and output.

6502 processor with 6530 I/O device (memory mapped)

Port A data I/O register   PAD =$1700
Port A direction register   PADD =$1701
Port B data I/O register   PBD =$1702
Port B direction register   PBDD =$1703

```
0200    A9 FF       LDA#$FF
0202    8D 01 17    STA PADD  ; Port A output
0205    A9 00       LDA#$00
0207    8D 03 17    STA PBDD  ; Port B input
020A    AD 02 17    LDA PBD   ; Read input
020D    C9 Read               ; Invert input
020E    8D 00 17    STA PAD   ; Display output
0211    4C 0A 02    JMP Read  ; Repeat
```

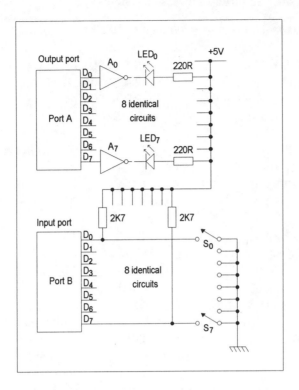

Figure 9.22   Interface for parallel to parallel data transfer

Z80 processor with non-memory mapped Z80 PIO. Note: additionally Port B strobe line STB should be set to logic 0.

Port A data I/O register   PAD = 04
Port A control register   PAC = 06
Port B data I/O register   PBD = 05
Port B control register   PBC = 07

0C90	3E 4F	LD,A 4F	
0C92	D3 07	OUT,(PBC)A	; Port B input
0C94	3E 0F	LDA 0F	
0C96	D3 06	OUT(PAC)A	; Port A output
0C98	DB 05	Read IN B(PBD)	; Read from Port B
0C9A	2F	CPL	; Invert byte
0C9B	D3 04	OUT (PAD)A	; Output to Port A
0C9D	18 F9	JR Read	; Repeat

6800 processor with memory mapped 6820 PIA.

Port A data direction register	DRA $8004
Port A control register	CRA $8005
Port B data direction register	DRB $8006
Port B control register	CRB $8007

0010	4F	CLR A	;Set up I/O registers
0101	B7 80 07	STA A CRB	"
0104	B7 80 05	STA A CRA	"
0107	B7 80 06	STA A DRB	"
010A	43	COM A	"
010B	B7 80 04	STA A DRA	"
010E	86 04	LDA A#$04	"
0110	B7 80 05	STA A CRA	"
0113	B7 80 07	STA A CRB	"
0116	B6 80 06 Read	LDA A DRB	
0119	43	COM A	;Invert data byte
011A	B7 80 04	STA A DRA	
011D	20 F7	BRA Read	;Repeat

A suite of programs designed to exercise many of the I/O devices described above is available from: Michael Tooley, BA, Dean of the Faculty of Technology, Brooklands College, Heath Road, Weybridge, Surrey KT13 8TT.

Test Questions

1. Calculate the approximate frequency of a 3-inverter ring oscillator when each stage has a propagation delay of 33ns. Sketch the circuit and draw one cycle of waveform showing the time scale and logic levels.
2. Sketch a circuit to show how two NAND gates can be used to debounce a two way switch.
3. (a) With reference to a floppy disk controller, state how write protection is achieved.
    (b) Explain how the data stream and control bits are organised within the tracks and sectors of a floppy disk.
    (c) State the reasons for formatting a disk before use.
    (d) State why it is necessary to detect track 00 during start-up.
4. (a) Explain what is meant by the term printer emulation.
    (b) Explain the significance of unidirectional and bidirectional printing.
    (c) Explain how the character F is produced by a dot matrix printer.
    (d) State the important functions of a floppy disk controller (FDC).

10.   Networks and transmission lines

Syllabus sections: B1–D.4, B1–E.5, B2–B.3, B2–B.4, M09–9.1, M09–9.2, I1.3, I2.2, I2.4, I3.5, B2–A.1, B2–A.2, B2–C.5, B2–C.6, B2–D.7.

Until the introduction of the computer into communications in the 1970s, typical systems were designed around the telephony networks on the PTT's (the national Post, Telegraph and Telephony organisations) public switched telephone networks (PSTN) and radio services. The PTT's networks connect individual subscribers to a second specified or addressed subscriber, so that this service is often described as *narrow-casting*. Whilst the radio services might well be intended for a specified receiver, the very nature of the transmission precludes specific addressing, so that the term *broadcasting* applies.

The communication of a computer with its peripheral devices forms the basic or primeval network with messages being passed over the data bus under the control of signals on the address and control buses. At a higher level, all the computer associated elements on a particular site may be interconnected to form a *local area network* (LAN). By using wired or radio carrier services, this concept can be expanded to cover an area as large as 50km diameter. This gives rise to the term, *metropolitan area network* (MAN). Further expansion to make use of international radio and PTT networks, form a *wide area* (WAN) or *global network*. The actual description of a LAN, MAN and WAN is not easy to define as one definition tends to merge into another. At the basic level, communications is often at baseband, thus forming a narrow band system. However, many popular

networks are broadband and use radio frequency multiplexing to increase the through-put of data. Networking requires the use of a modem to handle the data transfer. Different types of networks may be interconnected by using *gateways* placed at strategic points.

The physical communications medium varies from twisted pair or coaxial copper cables, through radio links to optical fibres. The last has the particular advantages of freedom from interference and electromagnetic effects and is capable of operating in very broadband networks. Networking makes better use of expensive peripheral devices, as these can be time shared between many users. The technique also increases the rate at which information can be disseminated within an organisation.

Network topologies

Bus or highway systems

All the communicating elements of the system are connected to a common cable in the manner shown in Figure 10.1. Each node has a unique address and only responds to data that is sent to it. In network terminology, a node is defined as a connection, tap or branch point. As the cable acts as a transmission line, its two ends must be suitably loaded to prevent signal reflections, and because it carries no power, it is said to be inactive.

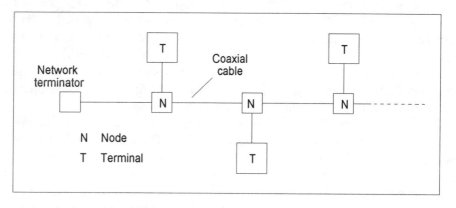

Figure 10.1   Bus or highway network

The network is controlled by a common clock which must be positioned near to the middle of the cable span. Its frequency then controls the rate of the data transfers. If the network length is increased beyond some critical value, then the clock rate has to be reduced to maintain an acceptable bit error rate (BER). Each node is allowed direct access to the network, but in contention with all the other devices. Each transmitter device listens continually to the network. If a collision of data is detected, then transmission ceases and the control protocol takes over. It is usually easy to extend this type of network, or to add or remove a terminal without interrupting the data flow.

Ring network systems

As the name implies, all the communicating devices are connected in a *ring*, with each device or terminal being equipped with an access node that has a unique address in the manner shown in Figure 10.2. Often no one node is responsible for network control and each has equal status. Transmission is normally uni-directional, although some bi-directional networks are in use.

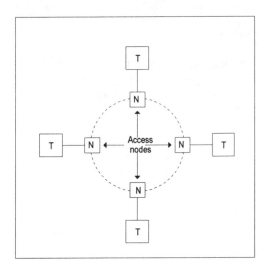

Figure 10.2   Ring network

The ring usually carries power for the node interfaces and each node acts as a data regenerator, receiving the data on one side and re-transmitting it to the next node in sequence. If the ring is broken either under fault conditions or to modify the terminals, then the data flow is interrupted. *Dual* and *braided* rings which have two parallel signal paths, can be used to overcome this problem.

Star network systems

Unlike the bus and ring networks, all the message switching and control is carried out at the central hub, as indicated in Figure 10.3.

173

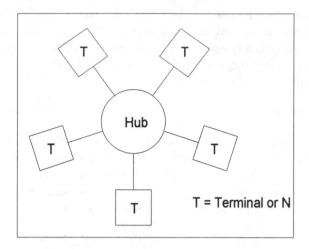

Figure 10.3   Star network

Since terminals can only communicate with each other via the hub, the configuration only finds favour where the data signals flow chiefly in one direction. Such an application would be the distribution of a television service.

Clusters

This is the term used to describe the terminal configuration where several peripheral devices are connected to the single drop cable from a node. The cluster will normally consist of a computer plus several peripherals such as printer, plotter, video display unit, keyboard etc. Control of the node is affected via the computer.

Value added networks (VANs)

This term is used to describe the additional value of services that may be provided over telecommunications networks in addition to the normal basic voice and data services. For example, Viewdata provides subscribers with access to public databases in addition to the normal telephonic services.

Network protocols

Contention techniques

This allows each terminal exclusive use of a network, but in contention with all the other terminals under the control of a technique known as *carrier sense-*

multiple access (CSMA). Carrier sense means that each terminal continually listens to the network to detect the absence or presence of a carrier. Multiple access is achieved because all terminals share the same transmission medium. Thus a device will not transmit its own data until it detects a gap in the transmissions. It is possible that two terminals waiting to transmit, both detect the same gap and start to transmit together and cause a data collision. When this happens, both transmitters halt and *back-off* for a time before trying to re-transmit. The control of this back-off period varies from system to system.

CSMA with collision avoidance (CSMA-CA)

This is based on an empty time slot system. Each node on the network is given a priority listing from a controlling node. If it has data to transmit, it does so only at its allocated time slot. If it has no data available, then this time slot is reallocated to the node with the next highest priority. Extra time slots can be allocated to those nodes that have a large amount of data to transmit.

CSMA with collision detection (CSMA-CD)

Data collisions can only occur during the short period after a transmission has started and before the signal has had time to reach all the other nodes on the network. This is the *collision window*. When a collision occurs, transmissions continue until all the nodes on the network recognise the burst of noise as a collision. The back-off period varies from a random interval, to a period calculated from the node address. Thus all terminals will back-off for different lengths of time.

Network systems

Ethernet system

This is a half-duplex send-receive system based on a highway or bus. It is constructed from a length of 50 ohms coaxial cable which must be terminated at each end in a suitable non-reflecting load. The typical maximum cable length is 2.5km, but this can be extended by interfacing other similar networks via suitable gateways. The maximum number of nodes or stations permitted on each sector is 1,024. The raw data rate is 10Mbit/sec and each node on the network is driven by a separate 20MHz clock.

The Manchester code format is used to ensure that the clocks are synchronised to the data stream. The bit cell is divided into two, the second part containing the true bit value and the first its complement. Network access control uses the *carrier sense, multiple access/collision detection* technique. The Thin Ethernet or Cheapernet system, operates at the same speed and in the same modes but uses the thinner 75 ohms coaxial cable. This restricts operations to a shorter network.

Cambridge ring system

This network, constructed from two pairs of twisted telephone pair cables, functions on the empty slot technique. Typically four slots circulate unidirectionally around the network, being passed between repeaters placed at each node. As these are actively connected to the ring for the whole of the operational time, the necessary power is supplied over the ring. One of the nodes is allocated to a fixed monitor station which creates the empty slots at start up, clears corrupted data packets and manages the error reporting function.

The success of this type of ring depends upon the cable characteristics. Any discrepancy in the lengths of the four wires can introduce a signal phase difference; the self capacitance and attenuation can cause distortion, all leading to an increase in bit errors. To counter these problems, the repeaters are usually spaced at intervals of less than 100 metres. The raw data rate around the ring is 10Mbit/s, but due to the addition of house-keeping bits etc., the actual rate is nearer to 1Mbit/s. The basic data coding depends upon the voltage transitions on the cable pairs, which nominally carry a dc potential of 28 volts.

A logic 1 is signified by a change of state on both pairs of wires, whilst a logic 0 produces a change of state on one pair only. For a succession of zeros, the state changes alternate between pairs. The data is organised into packets each of 40 bits. These contain data bytes, source and destination addresses, response and control bits. Each circulating slot carries one packet.

The address space for each node occupies 8 bits allowing for 256 possible network stations. The all-zero pattern is reserved for the monitor and the all ones indicate a packet that is broadcast to all stations.

Fibre distributed data interface (FDDI) network

This network, which operates as a *token* ring, uses optical fibre as the transmission carrier. It has been designed so that it can be coupled through suitable gateways into either Ethernet or a standard ring network. A *token* is a specific bit pattern within a data packet which circulates around the ring. A node waits until it receives such a packet, then removes the token and inserts its own data. The modified packet then circulates to its destination where it is removed. The receiving node then sets a flag bit in the packet and returns it to the sender. Reference to the flag bit signifies reception.

The carrier wavelength is typically 1300nm and this can easily accommodate the bit rate of 100Mbit/s. Rings of greater than 100km circumference can be constructed, and with up to 500 nodes per ring. The network consists of two separate fibres which connect to each node. This allows the network to be reconfigured under fault conditions at any node. Unlike normal token rings, the FDDI ring node releases the token immediately it has transmitted its data. This utilises the ring bandwidth and time more effectively so that many messages can be circulating simultaneously.

An FDDI-2 system has been designed to expand the flexibility of the concept and cater for both packet and circuit switched communications. This supports the 64Kbit/s circuit switched voice channels of an ISDN system. Of the total 100 Mbit/s rate, more than 98Mbit/s is available to share between packet and circuit switched applications. The remaining bit rate is used for synchronisation and packet delimiting.

A further expansion of this concept, FDVDI (fibre distributed video/voice data interface) integrates video signals with the data and voice services using a gross data rate of 2.48Gbit/s.

OSI (open systems interconnect)

This concept has been developed to enable dissimilar and normally incompatible devices, to exchange data by means of an agreed set of protocols. It is based on a defined system of interconnections and interactions between the seven functional layers or levels of the model shown in Figure 10.4.

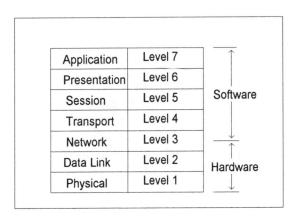

Figure 10.4   OSI 7-layer model

The layers below level 3 are related to system hardware and electrical interfaces, whilst the layers above are technology independent and related mainly to software. Interaction between terminals occurs at the same level, i.e. level 3 to level 3.

177

Network security

The most serious affliction for any networked system is the intrusion by hackers. At best, this represents an illegal access to secret files and sensitive databases, but more importantly, it provides an opportunity for the theft of corporate information and money. In addition, the vindictive and malicious hacker may also introduce a virus that can create havoc and confusion throughout a data base. It is therefore important that the network of any computerised database should be adequately protected to prevent any unauthorised access.

The *call-back* system provides a simple form of control. Each user has to make a call to the access control and provide a password. This is checked against a library listing to verify the password and ascertain the permitted level of access. If the caller fails to meet the required criteria, control terminates the call. If the caller is recognised as legitimate, he/she will be asked to terminate the call, where upon, control will call-back to a pre-arranged address number to establish the communications link.

Physical keys and passwords alone form a poor defence because both may be easily bypassed by a determined attacker. In the most secure system, the data will be stored in an encrypted form.

For telephone line based systems, leased lines are more secure than dial-up lines because these simply connect two specified locations. In a similar way, private packet based systems are more secure than public PTT networks. Local area networks (LANs) are the most vulnerable to attack and should be protected by the use of encryption, passwords and physical keys.

Bus extension systems

The central processor unit within a computer utilises a number of parallel path networks for its own communication and control. These paths which cater for *data, addresses and control* are referred as the *bus system*. Extending these buses into the outside world, expands the range of influence of the processor. In general, this is achieved by constructing the computer as a series of plug in cards mounted in a *back-plane*.

Two basic technologies are involved using either synchronous or asynchronous data transfers. With synchronous operation, the instants of data transfer are controlled by a system clock. By comparison, asynchronous systems may transfer one data package immediately after the previous one without waiting for the tick of a clock.

Bus extension concepts developed from a need to integrate various modular electronic devices into a computer controlled environment, modular automated test equipment (MATE) being a particular influence. The rate of data transfers is governed by the bus length so that increasing the area of influence reduces the data throughput. Variations on this basic concept has caused many *de facto* standards to be developed, some of which have achieved international standards status.

General purpose interface bus (GPIB)

This system is also known as the IEEE-488 or Hewlett Packard Interface Bus (HPIB) after the originators. The bus was designed specifically for instrumentation purposes and consists of eight parallel, bi-directional data lines, three control lines, plus five lines for general interface control. Up to 15 devices can be attached to the bus at any one time, but the total transmission line length must not exceed 20 metres.

Table 10.1   GPIB Interface

Pin	Signal	Pin	Signal
1	Data I/O 1	13	Data I/O 5
2	Data I/O 2	14	Data I/O 6
3	Data I/O 3	15	Data I/O 7
4	Data I/O 4	16	Data I/O 8
5	EOI	17	REN
6	DAV	18	DAV ground
7	NRFD	19	NRFD ground
8	NDAC	20	NDAC ground
9	IFC	21	IFC ground
10	SRQ	22	SRQ ground
11	ATN	23	ATN ground
12	Shield	24	Logic ground

Key:

ATN Attention.	NDAC Not data accepted.
DAV Data valid.	NFRD Not ready for data.
EOI End or identify.	REN Remote enable.
IFC Interface clear.	SRQ Service request.

TTL logic levels are employed, with logic 0 (High) between 2.5 and 5 volts and logic 1 (Low) between -0.6 and 0.8 volts.

Three types of devices may be connected to the bus and these are described as either, *talkers*, *listeners*, or *controllers*. For example, a voltmeter may be described as a talker, a printer as a listener, and a computer as a talker, listener or a controller. Although the network may contain more than one controller, one is

179

designated as master and allocates the bus to individual devices in turn. The system uses a standard 24 pin connector and cable as shown in Table 10.1.

Data transfer control (handshake)

The network can support several active listeners and these may not all be able to accept data at the same rate. Thus the slowest listener controls the rate of data transfers. The relative effect of the three control line signals, DAV, NRFD and NDAC are shown in Figure 10.5.

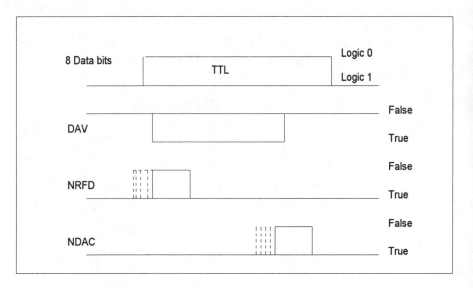

Figure 10.5   IEEE-488 handshake/control signals

NRFD line Active listeners use this line to signal their status with regards availability to accept data. Each device holds this line Low until it can accept data. The active talker now waits until all listeners have released this control line (now in the High state) before transmitting the next byte of data.

DAV line After placing valid data on the data lines, the active talker pulls the DAV line Low, signalling to the listeners that the data can now be read. A 2ms period is allowed for the data lines to settle to valid logic levels. All listeners now respond by pulling the NRFD line Low.

NDAC line With the DAV line Low, active listeners accept the data and then release their hold on the NDAC line which was previously held Low. When all the listeners have released hold on the NDAC line, it automatically returns to its High state. This signifies that all listeners have accepted the data byte. Data transfer is ended by the talker releasing the DAV line. This causes the listeners to pull down the NDAC line ready for the next data transfer.

180

ATN line (Attention) When this line is asserted or taken Low, any active talker releases control at the DAV line. The active controller then transmits information in the same way as other data byte transfers. Table 10.2 shows how all devices accept the data that is transmitted when ATN is asserted.

*IFC line (Interface clear)*This is used by the active controller to override all other activity.

SRQ line (Service request) This line is asserted by any device that requires service.

Table 10.2   Control data bit functions

Type	b_7	b_6	b_5	b_4	b_3	b_3	b_2	b_1	b_0
Bus command	x	0	0	C	C	C	C	C	C
Listen address	x	0	1	L	L	L	L	L	L
Talk address	x	1	0	T	T	T	T	T	T
Secondary address	x	1	1	S	S	S	S	S	S

Notes

b_7 is represented by a Don't care state x, whilst b_6 and b_5 are used to classify the type of command.

A listener receiving its address becomes active for byte starting $x01$.

A talker receiving its address becomes active for byte starting $x10$.

Address 31 (11111) causes listeners to become inactive. This is described as the *unlisten and untalk* address.

Secondary addresses for bytes starting $x11$ are intended for addresses of sub-units within a device.

REN line (Remote enable) This is used to disable the controls of a device. When REN is High the device reverts to manual control.

EOI (End or identity) When asserted by an active talker this indicates the end of message. When asserted in conjunction with ATN by an active controller, this calls for each device to identify itself and its status.

Hewlett Packard interface loop (HPIL)

Many serial data driven, low input impedance devices need to be driven from a current source. Two common loop current values are in use, either 20 or 60mA, with a logic 1 or 0 being represented by the presence or absence of the current. The HPIL is one proprietary standard that has been developed to handle medium speed, portable instrumentation requirements.

The communications link consists of a two wire serial loop connecting each device to a controller. Messages are transmitted in blocks of 11 bits, 8 bits of data, plus 3 control bits. These last define the way in which a device must operate before it can send data over the network.

I^2C bus (inter-IC)

This represents a small standardised 2 wire bus system configured as a *ring* network and devised for the inter-communication between integrated circuits. It can be adapted for use in the control of analogue or digital systems, ranging from a television receiver to industrial applications. Because only two lines are dedicated to control, the cost in terms of additional pin-outs on each chip is small. The two lines provide for bidirectional serial data transfers up to a rate of 100 kbit/s plus a clock signal. The bus operates on a master/slave basis but allows for several devices to act in turn as master, each one providing its own clock signal.

An arbitration procedure prevents bus-conflict or contention between devices, thus allowing only one master to gain control at any one time. Each IC on the bus has a unique 7-bit address which is contained in the first byte of each data transfer. Each device on the bus compares this byte with its own address and only if a match is found will it respond to the master that transmitted the address. The number of data bytes that follow is unrestricted and the slave acknowledges the received data on a byte by byte basis. Due to its self capacitance the bus is limited to a length of about 3.5 metres. Provision is also made to allow for the use of devices of different semiconductor technologies, for example, C-MOS, bipolar etc.

Small computer systems interface (SCSI)

This network facility, pronounced Scuzzy in the jargon, is a standard interface bus that allows computers to communicate with any peripheral device that carries embedded intelligence. The standard is covered by the American National Standards Institute (ANSI) and has developed into SCSI-2. Different types of device can be connected in a daisy chain fashion via a common 50-way cable, both ends of which must be correctly loaded in a suitable impedance. All signals on the cable are thus common to all devices.

To avoid bus contention, each device connected to the bus is given a unique 'SCSI' address with each address being allocated a degree of priority. The bus carries 8-bit data, 1-bit parity and 9-bit control lines, to provide a maximum synchronous data transfer rate of 5Mbytes/s. The maximum bus length is dependent upon the type of bus driver/receiver used. For single-ended devices, the length is restricted to 6 metres. This can be extended to 25 metres through the use of differential driver/receivers. SCSI-2 has been extended to increase the data rate to 10Mbytes/s with either 16- or 32-bit processors. The maximum data rate is thus nearer to 40Mbytes/s. The protocol has also been expanded to include *tagged commands*. This allows the execution of queued control commands according to a prescribed sequence.

PCMCIA (personal computer memory card (interface) international association

This concept developed from the need to provide memory expansion for the small portable, lap-top IBM compatible computers. The hardware electronics is contained within a credit card sized package which is plugged into a slot in the case side. Single or multi-layer printed circuit boards are carried in a plastic frame with stainless steel top and bottom covers. Mechanically this provides a very rugged structure with a small outline.

The original Type I slot has a height of 3.3mm, but because of the industry needs, this rapidly developed into Type II at 5.0mm, and Type III at 10.5mm heights. The electronic circuitry provides for high speed operation with low power consumption. The Type II interface allows the portable computer to be connected to a wide range of devices, such as LANs, small video cameras, satellite navigation receivers, fax machine, modems, apart from the original concept of memory expansion. The higher Type III slot can accommodate a low height disk drive or two slimmer cards.

Memory expansion caters for virtually all memory technologies, one time programmable ROM, SRAM, DRAM, EEPROM, flash memory, etc., with battery back-up to cater for non-volatile memories. The memory capacity extends up to 64Mbytes standard or 80Mbytes with compression. A standard 68-pin interconnection provides for either an 8- or 16-bit data bus and up to 23 bits for the address bus. Any unused address pins are left unconnected (NC). Data transfer rates in excess of 1.2Mbytes/s can be achieved with fast access (better than 200ns) to memory.

An unusual feature of PCMCIA is that it allows hot insertion or extraction of the module without the need to re-boot the computer. This has given rise to the jargon terminology, *Plug'n'Play* (P-n-P).

Centronics interface

This is a commonly used parallel interconnection, being originally designed by Centronics Corp. for use with their printers. It is based on a 36-pin Amphenol type connector and cable in the manner shown in Table 10.3. The interface provides for 8 bits of data, 2 handshake lines, a number of special control lines. If more than one printer is needed simultaneously, each must be driven from a further parallel interface.

The Data and Strobe lines are driven from the interface via the computer whilst the Acknowledge (ACK) line is driven from the printer. Data is placed on the 8 data lines and the Strobe line driven Low to transfer the data to the printer memory. The printer responds with the ACK pulse if the byte has been accepted. See Figure 10.6 for a diagram of these timing signals.

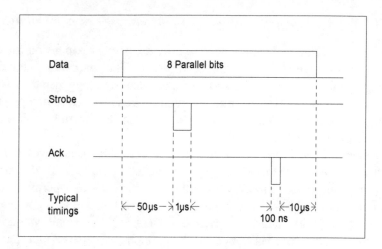

Figure 10.6   Centronics interface control signals and timing

Table 10.3   Centronics interface

Signal pin	Return pin	Signal
1	19	Data strobe
2	20	Data bit 1
3	21	Data bit 2
4	22	Data bit 3
5	23	Data bit 4
6	24	Data bit 5
7	25	Data bit 6
8	26	Data bit 7
9	27	Data bit 8
10	28	Acknowledge
11	29	Busy
12	30	Paper end
13		Select
	31	Input prime
14		Supply ground
	32	Fault
15		Osctx
16		Logic ground
17		Chassis ground
18		+5 volts DC

Functions of other lines:

Busy — High if printer cannot accept data.
Paper end — High indicates an out-of-paper condition.
Select — High indicates that this printer has been selected.
Input prime — A Low pulse from the interface resets the printer.
Fault — Low level indicates that printer has developed a fault.

Note

Pins 12, 13, 14, 15, 18, 31, 32, 34, 35 and 36 vary in function. They are commonly used for printer auxiliary control and error reporting. The cable should consist of twisted pairs for the signal lines 1 to 12 and the corresponding ground returns.

Recommended standard (RS) interfaces

In order to simplify the problem of inter-connecting a wide range of devices, a number of standard interconnections have been defined for various applications.

RS-232C

A well established standard for serial communication path lengths of less than about 15 metres and data rates below 20Kbit/s. However, using special modulation techniques, up to 76.8kBauds can be achieved. The interface is based on a 25-pin connector and cable in the manner shown in Table 10.4 and is designed for the interconnection between 2 devices. These are referred to as data terminal equipment (DTE) and data communications equipment (DCE). In general, DTE is the data originating device and DCE the modem connected to the communications lines. In many applications, not all of the facilities are used, so that there are a significant number of variations within this standard, even different cables and connectors can be employed. Although still commonly referred to as RS-232C, this standard has since 1986, been covered by RS-232D. Table 10.4 shows the standard pin assignments of the RS-232 connector.

Figure 10.7 shows the relative voltage levels of the RS-232 signals in relationship to noise immunity bands. Note that these signals use negative logic.

Figure 10.8 shows the relationship between RS-232 signals and the standard TTL levels for the ASCII character 'G'.

Table 10.4   RS-232C Interface

Pin	Signal	Source	Key
1	–	–	Protective ground
2	TXD	DTE	Transmitted data
3	RXD	DCE	Received data
4	RTS	DTE	Request to send
5	CTS	DCE	Clear to send
6	DSR	DCE	Data set ready
7	–	–	Signal ground
8	DCD	DCE	Data carrier detect
9	–	–	Reserved for data set test
10	–	–	Reserved for data set test
11	–	–	Unassigned
12	–	DCE	Secondary received signal detector
13	–	DCE	Secondary clear to send
14	–	DTE	Secondary transmitted data
15	–	DCE	Transmission signal element timing
16	-	DCE	Secondary received data
17	–	DCE	Receiver signal element timing
18	–	–	Unassigned
19	–	DTE	Secondary request to send
20	DTR	DTE	Data terminal ready
21	–	DCE	Signal quality detector
22	–	DCE	Ring indicator
23	–	DTE/DCE	Data signal rate selector
24	–	DTE	Transmit signal element timing
25	–	–	Unassigned

Note: These signals are viewed from the DTE.

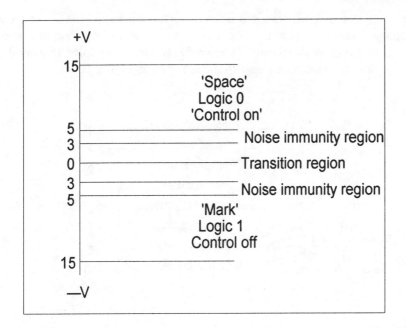

Figure 10.7   RS-232 voltage levels

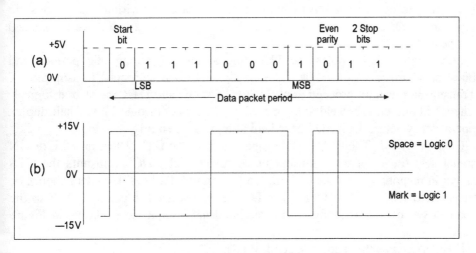

Figure 10.8   ASCII character 'G' symbol: (a) TTL logic level, (b) RS-232 signal

Figure 10.9 shows the standard interconnections between DTE and DCE. The secondary channel which is only rarely used (auxiliary communications) is identical to the primary channel. When a DTE is to be coupled to a modem, the wiring pattern is made on a 1-to-1 basis as shown in Figure 10.9.

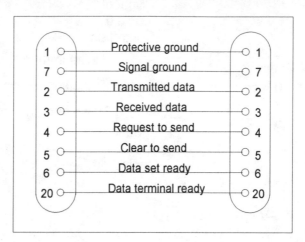

Figure 10.9   Important RS-232 link cabling

However, one DTE can be coupled to another via the RS-232 interface in what is referred to as a *null modem* condition. In this case, the wiring connections are again made on 1-to-1 basis, except that the cables to pins 2 and 3 are transposed at one end.

Data signals DTE holds the TXD line at logic 1 during idle periods and between characters (logic 1 is the Start signal for asynchronous transmissions). Transmission cannot proceed until RTS, CTS, DSR and DTR are also at logic 1. The RXD line must be held at logic 1 when DCD is Off (logic 1). For half duplex operation systems, the RXD line is held at logic 1 when RTS is at logic 0.

Control signals The RTS signal originates from the DTE where an OFF or ON signal sets the receive or transmit mode respectively. DCE transmits the CTS signal in response to an ON on both the DSR and RTS lines. (No data should be transmitted when CTS is OFF). The DTR signal is used to connect DCE to the channel when DTE is ready. The basic handshake sequence is shown in Figure 10.10.

DTR originates the request to send (RTS);

DCE responds with the clear to send (CTS);

DTE sets data set ready (DSR) to connect DCE to the channel;

DTE then transmits the data.

At the end of transmission, DTE takes RTS and DSR to OFF state, triggering DCE to set CTS to OFF.

The Signal quality detector line from DCE is held OFF only under conditions when errors are likely, otherwise it is held ON.

The data Signal rate detector line designates 1 of 2 data rates and is set ON to select the higher value.

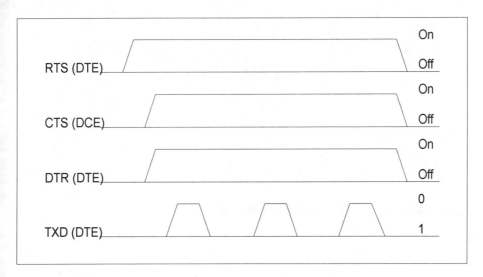

Figure 10.10   RS-232 control/handshake signals

RS-422 and 423

These standards were developed to overcome the low data rates and short line length problems associated with RS-232C. The maximum bit rates now are in the order of 10Mbit/s and 100Kbit/s respectively. The mechanical standard is generally the same as for RS-485 and RS-449 which is shown in Tables 10.5 and 10.6. Again, not all the facilities need be used in any particular application.

The performance improvements are achieved by using differential serial transmission and balanced driver/receiver amplifiers as shown in Figure 10.11 and Figure 10.12. RS-422 is a double balanced system, whilst RS-423 represents a balanced to unbalanced circuit to maintain compatibility with RS-232 systems.

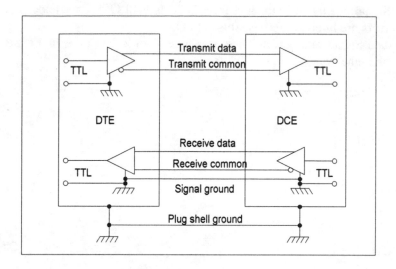

Figure 10.11   RS-422 full-duplex wiring

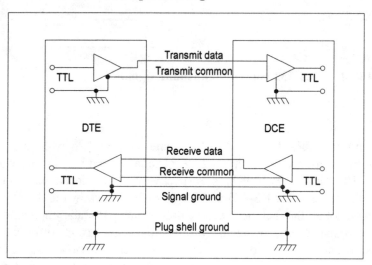

Figure 10.12   RS-423 full-duplex wiring

Table 10.5 RS-449 Interface

Pin	Signal	Pin	Signal
1	Shield	20	Receive common
2	Signalling rate indicator	21	Unassigned
3	Unassigned	22	Send data
4	Send data	23	Send timing
5	Send timing	24	Receive timing
6	Receive data	25	Request to send
7	Request to send	26	Receive timing
8	Receive timing	27	Clear to send
9	Clear to send	28	Terminal in service
10	Local loopback	29	Data mode
11	Data mode	30	Terminal ready
12	Terminal ready	31	Receiver ready
13	Receiver ready	32	Select standby
14	Remote loopback	33	Signal quality
15	Incoming call	34	New signal
16	Select frequency	35	Terminal timing
17	Terminal timing	36	Standby indicator
18	Test mode	37	Send common
19	Signal ground		

Note. In general, this table is also applicable to RS-422, RS-423 and RS-485 standards.

Table 10.6 RS-485 Standard Interface

Pin	Signal	Pin	Signal
1	Common Ground	6	CTS-B
2	RTS-A	7	CTS-A
3	RTS-B	8	Rx-B
4	Tx(Rx)-B	9	Rx-A
5	Tx(Rx)-A		

RS-530 Interface

This standard supersedes RS-449 and whilst it uses the same 25-pin connector system as RS-232, it is not pin compatible with it. RS-530 can be used with RS-422 (balanced) or RS-423 (unbalanced) interfaces and is also compatible with RS-449 and other interfaces. The standard provides for data rates up to 2Mbit/s.

Modems (MOdulators/DEModulators)

A modem is the device that is positioned between a data terminal and the transmission medium. The modem accepts data input in the serial format and then processes this to generate a form suitable for transmission. In the reverse direction, the modem performs a complementary function. For effective operation, the modem not only has to manage the communications facility, but also needs to be have an inbuilt control feature. At the basic level, this involves a form of hand-shaking between the calling and called terminals. This can create a problem with radio links because it is necessary to provide a reverse channel. The round trip propagation delay, particularly via satellite links, can be troublesome.

Each modem has to match the channel bandwidth and bit rates, provide synchronism for the data stream at both ends, handle the modulation and demodulation processes, and maintain the high signal to noise ratio needed to ensure the integrity of the data signal.

A wide range of modem types are manufactured to the recommended standards of the International Telecommunications Union (ITU) and the Comite Consultatif International Téléphonique et Télégraphique (CCITT). These range from early V21 devices running at 200 Bauds, through to Vfast modems running at a rate of 28Kbit/s. By using data compression, these later types can have an overall effective data rate of about 100Kbit/s. In general, all new modems tend to be backward compatible with earlier devices to ensure an economical upgrade for the end user.

Several different modulation techniques are employed, ranging from two tone frequency shift keying (FSK), through variants of phase shift keying (PSK), differential PSK (DPSK) to multi-level quadrature amplitude modulation (QAM).

Two *smart* techniques that are used in recently developed modems include *channel probing* and selective *repeat-request*. The former is used during the initial handshake period to probe the quality of the channel signal in order to select the optimum data rate. The latter technique is an error control protocol. If a number of data frames have been detected as being received in error, a single message can request the repeat of a given number of frames.

Modem data compression techniques

The bulk of the information handled by network modems is contained in text files. This means that there is a high probability that most of the symbols used will be contained within the standard 128 element ASCII (American Standard Code for Information Interchange) code set. Within this code set, shown in Table 10.7, there are a number of symbols that will have a high probability of occurrence, for example, the vowels a, e, i, o, or u. Each of these can then be represented by a

Table 10.7   ASCII code set

b_3	b_2	b_1	b_0	0	0	0	0	1	1	1	1	b_6
				0	0	1	1	0	0	1	1	b_5
				0	1	0	1	0	1	0	1	b_4
0	0	0	0	NUL	DLE	SP	0	@	P		p	
0	0	0	1	SOH	DC1	!	1	A	Q	a	q	
0	0	1	0	STX	DC2	"	2	B	R	b	r	
0	0	1	1	ETX	DC3	#	3	C	S	c	s	
0	1	0	0	EOT	DC4	$	4	D	T	d	t	
0	1	0	1	ENQ	NAK	%	5	E	U	e	u	
0	1	1	0	ACK	SYN	&	6	F	V	f	v	
0	1	1	1	BEL	ETB	'	7	G	W	g	w	
1	0	0	0	BS	CAN	(	8	H	X	h	x	
1	0	0	1	H	EM	)	9	I	Y	i	y	
1	0	1	0	L	SUB	*	:	J	Z	j	z	
1	0	1	1	VT	ESC	+	;	K	[	k	1	
1	0	0	0	FF	FS	,	<	L	/	l		
1	1	0	1	CR	GS	-	=	M	]	m		
1	1	1	0	SO	RS	.	>	N	^	n		
1	1	1	1	SI	US	/	?	O	o	DEL		

short code with a maximum of three bits, whilst the character z which has a low probability is allocated a full 8-bit code. This algorithm can be further refined by searching for long strings of characters such as *the, and, there,* etc., and then allocating short codes to these occurrences. The most commonly occurring words can then be held in an adaptively updated memory known as a *dictionary*.

Modem protocols and access control

The data to be transmitted is normally organised into packets by a device described as a *packet assembler/dissembler* (PAD). Each packet contains a *header*, a number of data bytes or octets, followed by a *footer*. The header can start with a series of synchronising bits, described as a preamble, followed by *destination* and *source* addresses. If the data block length is variable, then the header may also contain information about this. The footer carries either parity bits or *cyclic redundancy check* bits.

At the receiver, the header and footer are stripped off and the check sum recalculated. If this is in agreement with the transmitted check sum, then the data can be accepted as correct. Otherwise a *request for repeat transmission* can be made. *Smart* modems containing an embedded microprocessor can be used to maximise the data throughput under noisy conditions in an adaptive manner. The packet size may be varied according to the error rate. As the error rate rises, the packet length is reduced. Such modems are capable of operating at a range of bit rates. Starting at the highest rate, and automatically falling back to a lower rate under noisy conditions. Modems so equipped with inbuilt intelligence can be designed to operate with a personal computer. This obviates the need to adjust the interface by manually resetting switches when it is required to communicate with a modem with different parameters.

British Telecom's (BT) digital network services

BT provides a number of special network services for the business user and these are known variously as KiloStream, MegaStream and SatStream.

KiloStream

This provides for point to point communications over leased lines and operates using the PCM system. Customer data rates can range from 2.4 Kbit/s up to 64 Kbit/s, making the system ISDN compatible. This flexibility provides the customer with an economic share of the available bandwidth. The data bits are transmitted in a 6 + 2 frame with groups of 6 data bits being delimited by Start and Status bits.

MegaStream

This network operates at 2.048Mbit/s and provides the customer with point to point high speed data communications for high volume voice, data or video applications.

SatStream

This business service operates over the European Communications Satellite (ECS) system using single channel per carrier (SCPC) and frequency division multiplex (FDM). This is a broadcast service allowing communications throughout Europe using relatively small, low cost earth stations. Because of the broadcast nature of the service which makes it vulnerable to eavesdropping, the network provides encryption. The system is ISDN compatible and the customer is allocated bandwidth in blocks of 64Kbit/s up to a maximum of 2.048Mbit/s, thus providing for voice, data and video services. The system can operate in the point to point, point to multi-point, uni-directional, or multi-directional modes.

Transmission lines and communications links

Electromagnetic energy may be considered as being carried by a system of conducting cables, or as being guided by the surrounding electric and electromagnetic fields when it travels through the free space between the cables. The concept of free space has a further significance. Free space, for which a vacuum is a good approximation, forms a region which is used to define the absolute standard and physical constants for the characterisation of electromagnetic waves. The *permittivity* (ε), *permeability* (μ) and *velocity of light* (c) for free space are related by the expression $c = 1/\sqrt{(\varepsilon_0\mu_0)}$.

Transmission cables are designed to carry signals with the lowest level of attenuation and Figure 10.13 indicates the typical construction of two variants. Coaxial cable shown in Figure 10.13(a) is designed so that the outer sheath of braided copper can be grounded to act as a screen of the signal currents being carried on the inner conductor. This is described as being an unbalanced configuration.

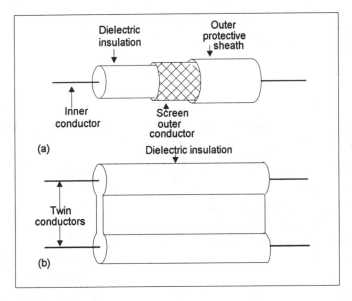

Figure 10.13   Typical transmission line construction: (a) coaxial cable, (b) parallel twin feeder cable

By comparison, parallel conductors may be supported by a dielectric material as indicated by Figure 10.13(b). In this case, neither line is directly earthed and the system is described as a balanced configuration because the signal currents flowing in the two lines, are always of equal magnitude but opposite phase. The

195

material used for the dielectric medium separating the two conductors in both cables for High Frequency (HF) signals upwards, ranges from air, a mixture of air and polythene to solid PTFE (poly-tetra-fluoro-ethylene). For lower frequency operations an unshielded twisted pair of wires (UTP) is often used.

For the purposes of transmission line analysis, it is convenient to consider the loop length of a transmission line system to be divided into two parts. One section considered to be ideal and lossless, whilst all the inductive, capacitive and resistive effects are referred to the other section in the manner shown in Figure 10.14(a).

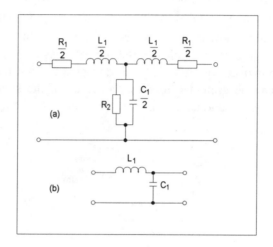

Figure 10.14   **Transmission line equivalent circuits: (a) the primary line constants, (b) the ideal lossless line**

This equivalent circuit represents the primary constants of the line in the following manner.

- Loop resistance R_1 per unit length,
- Loop inductance L_1 per unit length,
- Shunt capacitance C_1 per unit length and
- Shunt conductance G or leakage resistance R_2 per unit length (G = $1/R_2$).

Whilst the values of R_1, L_1, C_1 and G all increase with length, that of R_2 falls. If the resistive losses can be considered negligible as in the case of relatively short line lengths, the circuit can be re-drawn as shown in Figure 10.14(b). This is often referred to as the loss free or ideal line. Using both representations, a length of transmission line is thus shown to act as a low pass filter. If digital signals are passed over such a line as a series of pulses, this filtering action can remove much of the high frequency energy so that the pulses spread in the manner shown in Figure 10.15. The overlapped regions which represent inter-symbol interference

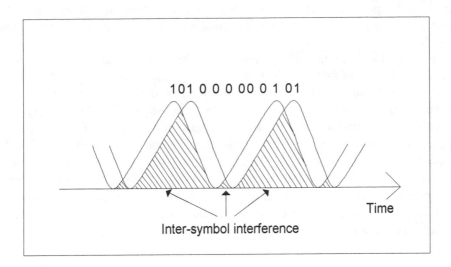

Figure 10.15   Pulse distortion showing inter-symbol interference

(ISI) then give rise to data errors at a distant receiver.

If the transmission line is driven by a sinusoidal signal, then the signal will propagate losing amplitude due to the resistance and acquire a relative phase shift due to the reactive components. The propagation constant thus has two components which both increase with line length.

Each line has a characteristic impedance (Z_0) that depends on the nature and dimensions of the materials from which it is constructed. This is defined as the input impedance of an infinitely long length of the line, or the input impedance of a shorter length when terminated in that value impedance. Z_0 can also be obtained from:

$$Z_0 = \sqrt{(Z_{sc} \times Z_{oc})}$$

where Z_{sc} and Z_{oc} are the input impedances when the line is terminated with a short circuit and an open circuit respectively. Z_0 values for cables commonly in use are 300Ω and 600Ω for parallel lines and 50Ω and 75Ω for coaxial cables.

Reflections and standing waves

When electromagnetic energy flows along a transmission line, the voltage and current distribution along the line, depends on the nature of the load at the far end. If the load impedance correctly matches the characteristic impedance of the line, then all the energy travelling on the line, will be absorbed by the load. Under mismatch conditions, the load can only absorb an amount of power that is dictated by Ohm's Law. Any surplus energy is reflected back along the line

197

towards the generator. The forward and reflected travelling waves combine to form *standing waves* all along the line, that are indicative of the degree of mismatch.

Figure 10.16 shows the voltage and current distribution near to the load for two extreme cases of mismatch, where the load is either an open or short circuit. In either case, there will be no power absorbed in the load, because either the current or the voltage is zero. Hence there will be a total reflection of energy. The standing wave pattern depends on the wavelength/frequency of the transmitted signal and is repetitive every half wavelength as shown by Figure 10.16. The impedance (V/I ratio) seen by the signal thus varies all along the line, being purely resistive of very high or very low value at the $\lambda/4$ points and either capacitive or inductive reactive in between. An open circuit at the end, behaves as a very low resistance just $\lambda/4$ away, whilst in the short circuit case it behaves in the opposite way.

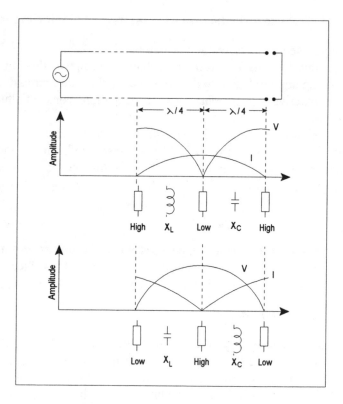

Figure 10.16   Mismatched transmission line

This impedance transformation property, allows transmission lines to be used for a number of other special purposes. A reflection coefficient is defined which is a function of the degree of mismatch between the load (Z_L) and line impedances and is given by $(Z_L - Z_o)/(Z_L + Z_o)$.

Artificial delay line Because electromagnetic energy takes a finite time to propagate along a transmission line, this feature can be used to construct a delay line. In a circuit such as shown in Fig.10.14(b), the time delay is given by $T_D = (L_1/C_1)$, where T_D is the time delay in seconds and L_1 and C_1 are the inductance and capacitance in Henrys and Farads respectively. For n identical sections in series, the total delay becomes nT_D.

Velocity ratio This is defined as the ratio of the velocity of electromagnetic wave propagation in a transmission line system to that which is obtained in free space. Values for typical cables range from about 0.6 to 0.85.

Practical cabling problems

The transference of data signals between two points can be basically accomplished by one of two methods. The data may be transmitted bit by bit over a single cable pair in a bit serial fashion or byte by byte over a set of parallel lines which is often referred to as a byte serial format. Obviously for equal bit rates, parallel transmission yields the highest data rate, but at a higher cost for cabling. For bit serial transmission, coaxial cable is able to support bit rates as high as 100MHz which is significantly greater than is possible over unshielded twisted pair (UTP) cable, but for only marginally higher cable costs.

For parallel transmission over multiway cables, there is a less obvious limitation. Over any significantly long length of cable, the bit timings across the medium can become staggered or skewed so that the bit rate errors rise. This effect increases in proportion to both cable length and bit rate. Parallel transmission is thus most effective if the data signals are contained in self clocking code format to aid synchronism at the receiver.

With serial transmission over a single cable pair, each byte of data can be delimited by a pair of Start and Stop bits to aid synchronism. This is referred to as asynchronous transmission. By comparison, synchronous transmission involves adding a special code pattern known as the clock synchronism word before each block of a number of data bytes.

Exercise 10.1

Study the ASCII code set and attempt to memorise the bit patterns for the alpha-numeric symbols.

Digital Techniques and Microprocessor Systems

Hints

The MSB for numerals is 0 whilst that for alphabetical symbols is 1.
The 4 LSB bits for numerals follows exactly the BCD code pattern.
The 3 MSB bits for numerals is 011.
Upper and lower case characters differ only in the bit 5 position. b5 = 0 for capitals and 1 for lower case.
Lower case characters start with 110 or 111.
Upper case characters start with 100 or 101.
All the alphabetical characters start from a = 110,0001 or A = 100,0001 and continue in alphabetical order.

Exercise 10.2

Inter-connect two computers via their RS-232 interface. Load a suitable software program and exchange files between the two data bases.

Exercise 10.3

With a signal generator set to about 100MHz and connected to a long length (say 100 metres) of coaxial cable, measure and calculate the characteristic impedance using the open circuit and short circuit, voltage and current values.

Exercise 10.4

Using the same set-up as for Exercise 3 and using a double beam oscilloscope, compare the phase and amplitude of the input and output waveforms. Use the same input signal levels but with the frequency set to 100MHz, 200MHz and 300MHz.

Test Questions

1.   With reference to networking, explain what is meant by the term 'cluster'.
2.   (a)   What are the significant advantages and disadvantages of using a networked system?
       (b)   What are the important features controlled by the network protocol?
3.   (a)   Describe the significant differences between ASK, FSK, PSK and QAM.
       (b)   What is the major advantage of QAM?
       (c)   Describe the important differences between 'baseband' and 'broadband' systems.
4   (a)   Explain why coaxial and twin cables are described as being 'unbalanced' and 'balanced' respectively.

(b)   Compare the advantages and disadvantages of serial and parallel communications with respect to data rates, cabling and costs.

(c)   State the 'threshold' and 'noise immunity' levels for the RS-232 system.

11　　Computer architecture

Syllabus reference: M02, I1, I2, B1, C3.

NOTE: You should remind yourself of the meaning of the terms bit, nibble, byte and word, see Chapter 2. Note also that the term Dword is also used to mean a double word of 32 bits. You should also remind yourself of the contents of the Chapter headed Microprocessors and Computing Systems in Volume 2, Part 1, of this series. The first part of this Chapter consists of a revision of these principles.

Definitions

In this chapter we shall not look in detail at the central processing unit or microprocessor because the important point at the moment is to see how it is used. The CPU or microprocessor chip will interact with all the other units of a computer system, and its actions are:
1.　　To read instruction codes from memory (ROM or RAM), and data from memory or other sources.
2.　　To act on the data as required by the instruction.
3.　　To output results to (RAM) memory, to other outputs, or to store temporarily for further actions.
Note that *CPU* can mean a set of ICs connected so as to provide the required actions, or it can mean a single chip, a microprocessor (*MPU*). The *architecture* of a computer means the way that the system is organised and constructed,

particularly with reference to the size of the data unit and the way that memory and peripherals are used. The *technology* refers to the type of MPU and associated chips in terms of chip construction (usually NMOS or CHMOS), and another important term is *word length*, meaning the number of bits handled as one unit. For modern computers, the Dword (32 bits) is the normal word length.

The *instruction codes* for a CPU can use one or more bytes per instruction, depending on how complex an action is. An instruction that requires all the bits in the main CPU register to be shifted one place to the right, for example, can consist of one byte. This byte is called an *operator*, it commands an action or operation.

If the action requires a byte to be loaded from memory into a register of the CPU, or the byte in memory to be changed, the instruction becomes more complex. As well as the operator portion, which might itself be of more than one byte, there must be an *operand*, one or more bytes which either provide the data directly or give an address in memory where the data is stored.

Data flow

A practical computer system will use both ROM and RAM memory, and provide for inputs and output. Inputs include keyboard, mouse and serial inputs from a modem. Outputs include video, printer, and serial outputs to a modem. The design of the system must allow for the easiest possible flow of data between all these units as required by the program instructions.

The instructions are held in program memory. This might be a separate chip, such as a ROM chip used only for program instructions, or it might simply be a portion of the RAM that contains a set of program instructions that have been read in from a disk.

The data bytes are held in data memory. This is usually a section of the main RAM using higher address numbers that are used for program bytes. Only a few items of fixed data are likely to be held in ROM and used when the machine is first switched on. The data that is to be used with a program will be loaded from a disk, typed at the keyboard, or input through a modem or other input device (mouse, scanner, CD-ROM drive etc.). Note that the RAM is used to hold both program and data bytes, and in many cases more of the RAM will be used for program than for data.

Inputs and outputs are a key portion of any computer system. The computer exists to process data, and the flow of data inputs and computed data outputs form the visible proof that the computer is working. The time that is needed to feed data in and out is critical, because it often accounts for most of the time that is needed for a computer to complete a task. By comparison, the time that the CPU spends on arithmetic and logic is often very small.

The CPU is normally heavily involved in inputs and outputs, because these are actions that are controlled by program instructions by way of the CPU, even when data entering through a port is stored directly in memory (see DMA, later in this Chapter). Some programs require very large amounts of computations as well as data-shifting actions, and such programs will run rather slowly on a conventional CPU which handles all such actions in sequence. Some machines can make use of a second processor (a *co-processor*) to handle computations, running in parallel with the main processor which is handling input and output actions.

All currently-available desktop machines use *serial processing*, meaning that the CPU deals with one instruction after another in sequence. The use of a co-processor is a simple example of parallel processing, and some large computers have been developed in which a large number of CPUs work in parallel so that all the portions of an action can be carried out simultaneously. Such parallel computers require a programming language (such as OCCAM) which is quite different from any used for conventional serial processing machines.

Bus structures

The conventional type of desktop computer relies heavily on the use of buses. A *bus* is a set of lines that make connections to several portions of the computer, so that data can flow between any pair of units that are connected to the same bus. This does not cause confusion, because the signals are gated to the bus lines under the control of the microprocessor, using the clock signals to determine when the gates open and close.

The use of buses avoids the problems that would otherwise arise in carrying data between any two of a large set of chips. If a bus system were not used, a set of input and output lines would need to be connected between each pair of units that could interchange data. For example, if a computer used CPU, memory, parallel port and serial port, full exchange of data would require six sets of input and output lines. By using a bus structure, each chip connects to the bus and the timing determines what signals are passing along the bus at any particular instant.

The disadvantage of a bus system is that it generally does not allow for data to be passed between more than two units simultaneously. All bus action is sequential; if you want data passed from unit A to units B and C you must use two steps, with data passed from A to B first and in another step from A (or B) to C. This also makes for slower action, so that a typical instruction might need several clock cycles simply to pass the data along the buses.

Buses are used for data, addresses and control signals. The *bus width* means the number of lines in a bus, so that an 8-bit bus width means that eight lines are used, each of which will transfer one bit of binary signal. For machine-tool computers, a 4-bit bus might be satisfactory, but desktop computers nowadays use a 32-bit (Dword) bus structure for both data and address. The early microcomputers used 16-bit address buses and 8-bit data buses. These were superseded by the 16-bit machines using 16-bit buses for both data and address, and after a short time by machines using larger address buses (20 lines, and later 32). The next generation of desktop computers will use 64-bit data buses.

NOTE: Buses are shown on block diagrams by wide arrows or by using an arrow with a slash line and a number to show the bus-width in bits. Both conventions are illustrated in this book. The arrows show the direction of data flow, and two-way arrows are used to indicate that data can flow either way.

Buses can be simple or *multiplexed*. Simple buses are more common for data, using one line for each bit. Multiplexed buses can be used for addressing, though they are not so common as they were. For example, the 8088 chip uses a multiplexed 20-bit address bus. The chip uses only 12 address pins, and these place the higher 12 bits of an address on the bus on one clock cycle. The other address bits use the same pins as the data bits, so that the lower 8 bits of an address are put out from the data pins on a different clock cycle. A simple octal latch chip such as 74LS373 can be used for demultiplexing such system, and the 74LS245 octal bus transceiver is also commonly used in such circuits. The advantage of using multiplexing is that it reduces the pin count on the chip, but the disadvantages of slower operation and the need for using a demultiplexer outweigh the advantage of a smaller pin number, particularly now that pin-grid arrays are used for CPUs rather than DIL packages.

NOTE: If multiplexed buses are used, the computer designer can decide at what point they will be de-multiplexed.

The three signal lines from the CPU labelled as S0 to S2 are used as inputs to a bus controller chip. The effect of this is to generate timing signals, so that at the time when an address is to be sent out, the 74LS373 chips are enabled to connect the data on A8–A19 and AD0–AD7 on to the 20 address lines. When data is to be input or output, the 74LS245 chip is enabled to allow the AD0–AD7 lines to be used for data.

Multiplexing was used on this chip and on others constructed at about the same time (late 1970s) so as to allow the standard 40-pin DIL chip package to be used. Later chip packages of 64-pin DIL, and the now universal PGA (Pin-grid array) of 130–160 pins arranged in a double row around a square, allow more pins to be used so that multiplexing like this is now unnecessary

Figure 11.1 illustrates the multiplexed system of the old 8088 CPU chip — this system was not used on the later chips such as the 80286, 80386, 80486 or

205

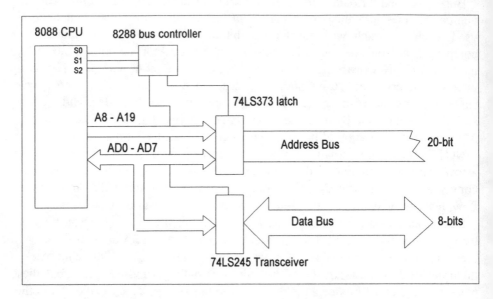

Figure 11.1 The multiplexed bus system of the Intel 8088 chip that was used in the original IBM PC machine in 1982

Pentium. The higher-order address pins A8 to A19 are shown as inputs to a latch — in practice several of these pins are also multiplexed with control signals, but that need not concern us now. The data lines AD0 to AD7 are also inputs to the latch. In practice more than one chip of this type would be needed because there are 20 lines in all, but since the same control signals are used it is easier to show this as one block in this diagram.

Standard system

Figure 11.2 shows the standard computer system illustration, to which the remainder of this Chapter will refer. This is a generalised 8-bit data system which would have been typical of small computers in the early 1980s, and its principles can still be applied to modern 32-bit machines. In this example, MPU is used in place of CPU to emphasise that the controller is a microprocessor.

The MPU is shown as placing signals on the address bus, because in a system like this with only one processor, the MPU has total control over the address bus — all other chips *receive* address signals. The data connection between the data bus and the MPU is two-way, because the MPU must read data as an input and also output processed data. The connection of the MPU to the control bus is also

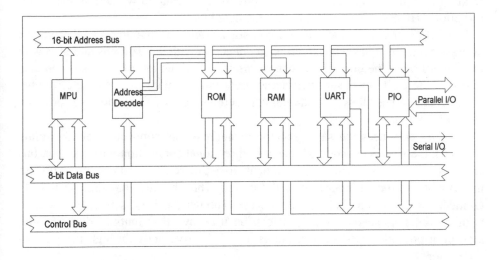

Figure 11.2   A simple standard system which demonstrates the features described in this chapter

two-way, because the MPU will issue some control signals to other chips and also receive signals from other chips. Note that ROM sends data to the data bus but does not receive data. You should memorise this diagram and the directions of the bus signals for each block.

The address *decoder* is the chip that allocates addresses. Some MPU address numbers must correspond to ROM and some to RAM, and a few are reserved for the UART (serial) and PIO (parallel) input/output chips. The address decoder chip receives signals from the address bus and from the control bus (originating in the MPU) and determines which of the four memory-using sections will be enabled. This chip prevents any possibility of contention for the use of the buses, because only the chip that is entitled to use the buses will be enabled. Chapter 4 dealt with address decoding for memory and other devices.

The *ROM* unit (usually one or two chips) receives signals from the address bus, and is enabled or disabled by an output from the address decoder. It is also controlled by control bus signals, and its output is placed on the data bus. This output will for the most part be program instructions to the MPU, such as start-up instructions, and short routines for such tasks as peripheral control (disk, VDU, keyboard etc.).

The *RAM* unit (which might consist of a large number of individual chips, each storing one bit of data) uses the signals from the address bus, and is enabled by the address decoder. It also receives signals from the control bus, notably the

207

read/write signals that determine the direction of data flow, and it has a bi-directional (two-way) connection with the data bus.

The *UART* chip uses the address bus signals and an enabling signal from the address decoder. It has a bi-directional connection with the data bus and also with the control bus. The control bus can determine the direction of data and can also interrupt the MPU to ensure that serial input is dealt with by a suitable routine. Two separate lines allow for external connections to RS-232 connectors for serial input and output.

The *PIO* chip uses the same scheme of address bus connection and enabling line from the address decoder. It has a bi-directional connection with the data bus and also with the control bus, so that it also can send signals to the MPU to interrupt it and run an input routine if required. This chip also has the parallel I/O connections to the parallel port connector. Though the PIO is usually operated with a printer as an output only, it can be used by other units bi-directionally, making it possible to connect scanners and disk drive units through the parallel connector.

Interrupts

An interrupt is an electrical signal to the MPU that will make the MPU stop whatever it is doing and run a program, a *service routine*, that will satisfy the needs of the interrupt. The MPU action then resumes where it left off. When an interrupt signal (which can be performed by software as well as by hardware) is received, the contents of the registers in the MPU are saved, using a piece of memory that is set aside for this purpose (by the programmer) and called the *stack*. The interrupt uses a code number, and this *interrupt number* determines which of a set of program routines will be used. For example, if the MPU is interrupted by the UART, the address of a routine for reading serial data will be put on the data bus, allowing the MPU to use this address and start the routine running to read in serial data.

Interrupts are of two types, *maskable* and *non-maskable*. The MPU uses separate pins for these signals, and the difference is important. A non-maskable interrupt cannot be deferred in any way; it must be obeyed at once. The usual candidate for this is a reset button, because only an emergency would call for such an interrupt. The maskable interrupt can be disabled by placing a 1 bit into a register inside the MPU (the *flag or status register*), and this can be done using software instructions. This is the way that one interrupt can be prevented from interrupting another, because when an interrupt routine starts running its first action is to set the mask bit to disable any other interrupt except the non-maskable type.

Older MPUs did not provide for a large number of types of interrupt, but modern computers can use an *interrupt controller* chip which will allocate priority if there is more than one interrupt pending, and which will also place the address number for the service routine on the bus.

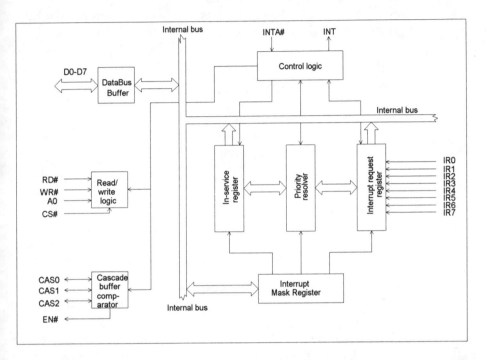

Figure 11.3   Block diagram of the 82C59A interrupt controller chip. The signal marked INTA# is the interrupt acknowledge signal from the MPU

An interrupt controller also allows several interrupt lines to be used rather than the single line for a maskable interrupt that is supplied on the MPU chip. Figure 11.3 illustrates the Intel 82C59A interrupt controller which can handle up to eight incoming interrupt signals, and which can be cascaded to allow up to 64 interrupt inputs.

The use of an interrupt can be illustrated by thinking about the use of a keyboard input. A computer must respond to the use of a keyboard, and the usual response will be either to carry out a command when a key is pressed, or to place the character on the screen and into memory when a key is pressed. At one time, this would be done by running a loop of commands.

A keyboard loop consists of a set of software program commands that test the keyboard to find if a key is depressed. If no key is found in this way, the command simply repeats, but when a key is found, the keyboard code is read and decoded to find which key has been used, and the character corresponding to that key is placed on the screen and also into memory.

This system, a *polling* system, Figure 11.4, is very wasteful because the MPU is idling at all times when a key is not pressed.

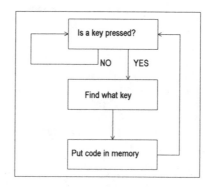

Figure 11.4   Illustrating a polling system or loop

The alternative is an *interrupt-driven* system. The MPU is free to carry out actions such as transferring data to memory or working out arithmetic until a key is pressed. The act of pressing any key causes an interrupt, and because this interrupt has come from the keyboard interface (not illustrated separately in the system diagram) it will cause the keyboard reading routine to run. The contents of the MPU registers are saved on the stack, and the keyboard is read, placing the character on screen and into memory. When this has been done, the register contents are replaced from the stack, and the MPU resumes its work. For example, in the course of a word-processing program, the screen may need to be updated as a result of reading the character from the keyboard.

The interrupt-driven system has the considerable advantage that it uses the MPU only when required. Even in an old and slow system with a 4MHz clock rate, the time between pressing two keys on the keyboard is very great compared with the clock time. Even if you can type at a speed of 4 characters per second, for example, this corresponds to 250ms time between interrupts, and at a (slow) 4MHz clock rate this allows for 1,000,000 clock pulses in the time between characters.

DMA

DMA is an acronym of direct memory access, and it refers to a system of transferring data without the need to read and write the MPU registers. The older computer designs of the 1970s used 8-bit microprocessors, and any actions that transferred bytes to and from memory was forced to make use of the MPU. For example, reading data from a disk was carried out by reading a byte of data from the disk to the accumulator register of the MPU, then writing the data into memory, and repeating this action for each byte that needed to be transferred.

This had two disadvantages, one that the speed of writing or reading was low because it involved two MPU actions, secondly that the MPU was tied up in repetitive actions. Basically the DMA chip can transfer bytes to and from memory and other devices, and the MPU is involved only in setting up the DMA chip (number of bytes, source of data, destination) and starting the process, leaving the MPU free to continue other work. The use of a DMA chip, for example, allows data transfers to be started by an interrupt signal.

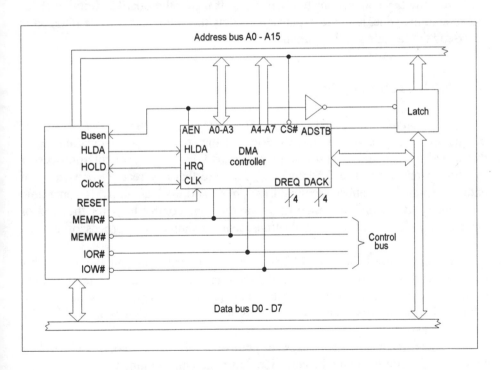

Figure 11.5 **The Intel 82C37A DMA controller being used in an older system based on the 8080 (similar to Z80) CPU**

211

In a typical memory-transfer action, the DMA chip works in three main steps. In the first step, the DMA chip registers are programmed with the address of the data source and the data destination and the number of bytes to be transferred. This step requires the use of the CPU. The next steps are reading and writing, which will take place, with incrementing of address numbers, until the byte count is zero. The action is slightly different when a port uses DMA, because the action is usually initiated by an interrupt signal and will then continue until no more data is available (there is no byte count used, and the action continues until the port buffer is empty). Figure 11.5 shows the block diagram of a DMA controller (Intel 82C37A) being used in an 8080 (8-bit) system. DMA is also known as *cycle stealing* because it halts the processor for the duration of a number of clock cycles.

Exercise 11.1

Examine a modern computer motherboard — the unit that contains the MPU, memory, ports and supporting chips. Identify the MPU, data bus, and address bus. Note the bus connections to other chips. If the motherboard is from an IBM-PC type of machine, note the expansion sockets that allow cards to be plugged in to extend the facilities of the computer.

The fetch-execute cycle

The *fetch-execute cycle* is the basic MPU set of steps that is performed for practically all actions that involve the use of memory, and an understanding of this cycle is essential if you are to understand bus timing in microprocessor systems of any type. The fundamental points are that any reading or writing of data must start by establishing an address on the address bus, allowing some time for memory to respond, and then using the data and control lines for the read or write action. This is most easily illustrated using a simple microprocessor such as the Z80.

Figure 11.6 shows the (simplified) timing of a read cycle on the Z80. This action extends over a set of four clock pulses, and in the middle of the logic-high state of the first of the pulses, the data bus is placed into a isolated state. On the falling edge of the first pulse, the address for the read action is placed on the address lines.

Since this microprocessor uses a comparatively slow clock by modern standards, the address could be considered as valid almost immediately.

The read control signal is then taken low in the middle of the logic-low state of the first pulse, so enabling reading. The data is not read until the falling slope of the second pulse, when the data bus is released and the signals become valid.

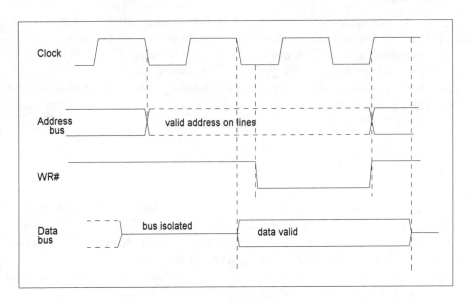

Figure 11.6   A simplified Z80 read cycle, showing the address and data bus signals, along with the RD# signal

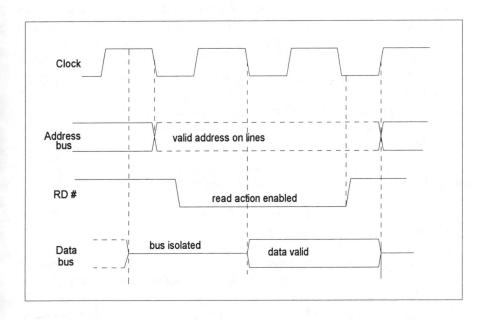

Figure 11.7   The write cycle for the Z80 buses

Figure 11.7 shows a Z80 write cycle. In this example, the address is again established first, on the trailing edge of the first pulse. Data is placed on the data bus at the end of the logic-high portion of the second pulse, and the WR# signals is taken low early in the logic-low portion of the second pulse.

The write action ends when the address bus control is ended at the leading edge of the fourth pulse, and the data ceases to be valid near the end of the logic-high portion of the fourth pulse.

Expansion buses

An expansion bus is a set of address, data and control bus lines that can be used by units other than the basic computer units. There are, in modern computers, several buses. The microprocessor itself will use internal buses that are organised to make data flow inside the MPU as fast as possible.

There will also be the buses on the main MPU board that deal with the main units or RAM, ROM, UART and PIO, and there may also be a 'local bus', a set of buffered bus lines that operate at the same high clock rate as is used by the RAM and ROM. Local buses are used to carry data to and from the disk controlled unit and the video controller unit

Finally there may be another buffered set of buses that run at a slower clock rate and which are used for add-on cards such as sound, multi-media, CD-ROM drive, extra serial ports and so on. These are the main expansion buses, and they permit the computer to be extended as and when new devices become available. The ability to extend easily is an essential feature of any computer intended for serious purposes, and it has been easy extension (and compatibility with older models) that has made the IBM PC type of machine pre-eminent.

Test Questions

Note: unless otherwise stated, these relate to 8-bit MPU systems.

1. An instruction consists of a 1-byte operator and a 2-byte operand. Suggest what the bytes of the operand might signify.
2. What is meant by a multiplexed bus?
3. When a byte is placed on the data bus by the MPU and written into memory, what prevents the same data being used at the same time by the PIO, UART and other chips connected to the bus?
4. What is the purpose of (a) a non-maskable interrupt, (b) a maskable interrupt?

5. What is the difference between a polled keyboard system and an interrupt-driven keyboard system?
6. What is DMA? What is the advantage of using DMA in a computer system?
7. In a typical fetch-execute cycle, why is there a time delay between placing an address on the address bus lines and performing an action such as read or write?
8. Name two of the signals that you would expect to find on a control bus.
9. Why are expansion buses usually run at a clock speed that is much lower than that of the MPU?
10. Name the blocks (referring to the standard system drawing) for which the following remarks are true:
   (a) The only block that outputs to the address bus.
   (b) The only block that receives signals only from the address bus and the control bus.
   (c) The only block that writes to the data bus but does not read from it.

12   Systems operation

Syllabus references: M03, I1, I3

The microprocessor

A microprocessor is a programmable logic chip which can make use of memory. The microprocessor can address memory, which means that it can select stored data, make use of it, and store results also in memory. Within the microprocessor chip itself, logic actions such as the standard NOT, AND, OR and XOR gate actions can be carried out on a set of bits, as well as a range of other register actions such as shift and rotate, and some simple arithmetic. The fact that any sequence of such actions can be carried out under the control of a program which is also read from memory is the final item that completes the definition of a microprocessor.

In general, microprocessors are designed so as to fall into one of two classes. One type is intended for industrial control, and this also extends to the control of domestic equipment, such as washing machines and central heating systems. A microprocessor of this type will often be almost completely self-contained, with its own memory built in, and very often this will include the programming instructions. Such microprocessors will very often need to work with a limited number of bits at a time, perhaps 4 or 8. The number of possible programming instructions need only be small, and clock speeds need not be high for most

216

applications. The control microprocessor will also be offered typically as a semi-custom device, with the programming instructions put in at the time of manufacture for one particular customer. Some older microprocessors originally intended for computing purposes, such as the 6502 and the Z80, are still being used extensively in controllers along with controllers which have been developed from these microprocessors and controllers which have been designed from scratch.

The alternative is the type of microprocessor whose main purpose is computing. These are of two types: the complex instruction set (CISC) and the reduced instruction set (RISC) chip types. Either type can be manufactured with little or no memory of its own, but each is capable of addressing large amounts of external memory and will deal with at least 16 bits, or, more usually, 32 bits of data at a time.

The CISC type has a much larger range of instructions, and will generally operate at fairly high speeds. The RISC type uses very few instructions, each of which can be executed very rapidly. The thinking behind RISC is that most microprocessors spend most of their working lives carrying out a relatively small number of instructions, so that by concentrating on the fast and efficient processing of these few instructions the processor will be faster. In practice, the need for more complicated actions to be carried out requires software for the RISC type of processor to be longer and more elaborate, reducing its advantages. Several types of computing processors now contain appreciable amounts of memory (16K typically) used as temporary storage (cache memory). This memory is fast (25ns or less) and is used to overcome the problem of slow main memory by reading or writing the main memory at a time when the microprocessor is otherwise occupied.

Figure 12.1 illustrates a generalised microprocessor unit (MPU) from around 1980, using an 8-bit data bus and a 16-bit address bus. Note that such microprocessors are still in use for computing purposes, notably in the Amstrad PCW machines and NC-200 notepad machines. This block diagram shows all the essentials that exist also in processors of later design, and understanding of this structure will make it much easier to understand the refinements that have been added in more recent designs.

The block diagram shows a number of registers, details of which we shall look at later, all connected by buses and control lines. There is one internal data bus which connects to the external data bus (the data pins of the MPU) through buffers that can be enabled as required. This allows the data flow within the MPU to be controlled independently of the data flow outside the MPU.

The timing of all actions is controlled by the clock, which nowadays is often a separate chip. The clock pulses are usually available externally, but in this diagram we look only at the internal action. The clock pulses are applied to the

217

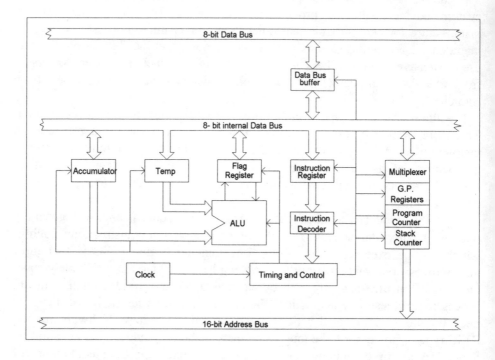

Figure 12.1   A block diagram for an 8-bit microprocessor suitable for machine-control use and for non-critical computing use

timing and control section, which in turn controls all the registers and buffers of the MPU.

Clocking

The clock pulse is the master timing pulse of any microprocessor system so that the specification of the clock pulse in terms of repetition rate, rise and fall times and pulse shape is fairly exacting. Some microprocessor chips include their own oscillator circuits, so that the only external components that need to be connected are a quartz crystal of the correct resonant frequency, and a few other discrete components.

More often, the clock is an external circuit which can be a chip supplied by the makers of the microprocessor, or a circuit constructed from logic gates of the TTL LS class. Figure 12.2 shows typical clock specifications for older Intel microprocessors working with clock periods of 200 to 500ns. The rise and fall

218

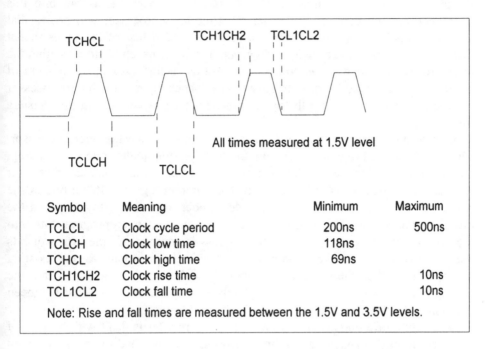

Note: Rise and fall times are measured between the 1.5V and 3.5V levels.

Figure 12.2 **The specification of the clock pulses for the older Intel processors 8088 and 8086**

times are particularly important, because slow rise and fall times can cause considerable timing problems in circuits that use microprocessors.

The specification of a minimum clock frequency (in terms of maximum clock period), which can be as high as 8MHz for a microprocessor that normally operates with a 12MHz clock, reflects the high leakage inside the chip, due to very close packing of tracks.

The requirement for maintaining fast rise and fall times for the clock pulse means that some care has to be taken with circuitry around the clock terminals of the microprocessor. This is not difficult if the clock circuits are built into the chip, or even if an external clock chip is used, because this can be located near the clock input pin of the microprocessor.

The instruction register

Each microprocessor chip contains one main control register which is designated as the *instruction register*. This will be of as many bits as the microprocessor is

219

designed to handle at a time, usually 8, 16 or 32. Whatever is put into this instruction register completely decides what the microprocessor will do, so that access to this register must be very severely limited. Most of the bytes that it deals with, in fact, come from a preset group that is permanently stored within the microprocessor, called the *microprogram*. By using this system, the makers of microprocessors avoid the need to have to check each input to the microprocessor in case it should contain conflicting commands, such as connecting all registers together.

A set of microprograms is stored in ROM memory inside the microprocessor. There will be one program for adding, another for subtracting, one for ANDing, another for ORing and so on. Putting a set of code bits into the instruction register will have the effect of calling up one of these microprograms. If the bits in the instruction register correspond to the code number for a microprogram, then the microprogram is run. This is done by feeding groups of microprogram bits into the instruction register in turn until the process is complete. If the group of bits that is used to call up the microprogram does not correspond to any existing microprogram, then the command is ignored. In this way, the gating within the microprocessor is controlled in a completely predictable way, one that has been determined by the manufacturers and is built into the chip.

The loading of a command into the instruction register is dealt with by part of the microprogramming. When the microprocessor is switched on, its first action will be to load in a byte, a word (16 bits) or double-word (32 bits, a Dword), according to whether it uses 8, 16 or 32-bit data units. This first byte, word or Dword is gated directly into the instruction register, and then these gates are disabled. If this code now calls up a microprogram, the instruction register is supplied from the internal microprogram, and gates will allow signals to be copied from one register to another or allow external signals from the pins of the chip to be connected to other internal registers following the instructions of the microprogram.

Following the last microprogram action, the gates which connect from the pins into the instruction register are enabled again so that another command word or Dword can be read. In this way, the microprocessor reads commands only when it is ready for them. Since each action is triggered by the arrival of a clock pulse, the timing is always exact, and each instruction will take a fixed (but different) number of clock pulses to run its microprogram.

Take, for example, the process of adding two numbers which we will imagine are already stored in two registers, with the result to be returned to one of the registers. The action of addition is started when the microprocessor reads an ADD instruction taken from the external memory. The ADD instruction is gated directly into the instruction register, which is then shut off as far as external signals are concerned. By analysing the bits of the instruction in the *instruction decoder*, the correct microprogram is called up. The first part of the micro-program is then loaded into the instruction register, and its action is to connect one register to one set of the inputs to an adder, a collection of gates, with two sets of inputs and one set of outputs.

This action requires one clock pulse, and on the next clock pulse the other register is connected to the other set of adder inputs. The next clock pulse provides the next microprogram instruction, which connects the (stored) output of the adder back to the input of one register. The next clock pulse brings in the microprogram action which enables the gates, so adding the bits and then storing the result back on the next clock pulse. The last microprogram action must then re-enable the gates which allow a command to enter the instruction register from outside the microprocessor.

This is oversimplified, particularly as regards the clock pulses, which are also gated to the correct registers. The principle, however, is that the actions in the ADD routine are decided by the microprogram which has been called into action by a single instruction. In addition, we have established the very important point that this single instruction reaches the instruction register only when gates are enabled, and that for most of the time the instruction register is not accessible to signals from outside the microprocessor. Another important principle is that all actions are carried out in sequence, one stage of action for each step in the microprogram. Finally, all of the actions are controlled by the clock pulses, and the speed of all processing depends on the speed of these clock signals.

Arithmetic and logic unit

Many of the instruction codes for the CPU will require arithmetic and logic actions, and these are dealt with by a part of the unit called the *arithmetic and logic unit (ALU)*. The ALU has two inputs, because most arithmetic and logic actions involve two numbers which are added, subtracted, multiplied, divided, ANDed, ORed and so on. The single output of the ALU is the result of the arithmetic or logic action. In practice, the CPU uses one main register, often called the *accumulator*, which provides one of the inputs to the ALU, and which also is used by the ALU to store an answer.

For example, referring to Figure 12.3, if the accumulator register stores the number 3FH (taking single byte actions for simplicity) and the CPU is executing an instruction that will AND the accumulator contents with 76H, then on the next clock cycle, the accumulator is connected to one input of the ALU and the byte 76H is connected to the other input. On the following clock cycle, the AND action is carried out and the result of 36H is loaded back into the accumulator, which now stores this byte in place of the 76H it held previously. Note that each

221

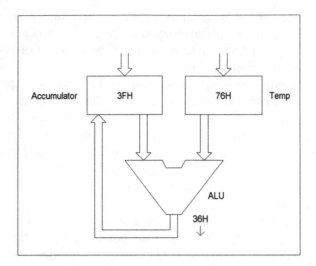

Figure 12.3   **The ALU action. For an action on two bytes, one is always taken from the accumulator and the result is returned to the accumulator**

instruction will require one or more clock cycles, and some actions will need a large number of clock cycles to complete.

The CPU requires a large number of switching and gating actions to be carried out. In the example above, the accumulator has to be connected to one input of the ALU and incoming data has to be connected to the other input. On the next cycle, the AND action is carried, using gating in the ALU, and subsequently the output of the ALU is gated into the accumulator, replacing the previous contents of the ALU. These switching and gating actions, as required by the instruction, are performed by a *control unit* within the CPU.

Accumulator, temp and flag registers

The *accumulator* is the main 'action' register of the MPU, and its output can be either to the ALU or to the internal data bus. In all actions where two items of data have to be combined (added, ANDed and so on), the accumulator will provide one byte, word or Dword and the register that has been marked *temp* will supply the other. The temporary register used in these operations is not controllable by program instructions.

The *flag register* (also known as the *status register*) exists to signal the results of actions, particularly actions in the accumulator, and each bit in this register signals a different result. The flag register can use 8, 16 or 32 bits according to the design of the processor, but not all of these bits will be used.

Figure 12.4 shows the layout of the Intel 8088 flag register, from which you can see that only 5 bits are used in a 16-bit register. In the following description the accumulator will be referred to, because earlier types of processors were constructed so that only actions in the accumulator affected the flag register. In later types, arithmetic and logic actions can be carried out in other registers and their results will affect the flag register.

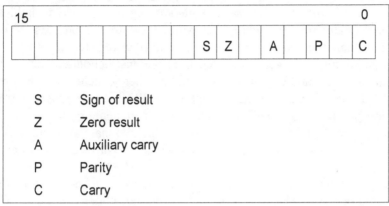

Figure 12.4   The flag register of the 8088 and the explanation of the lettering

The S flag bit is set if the most significant bit in the accumulator is set. If the program is working with signed (positive or negative) numbers, this indicates that the result is negative. The Z flag is set if the result in the accumulator is zero; this is particularly useful for checking if the data in the register is equal to some fixed data. The A flag is more specialised — it is set if there is a carry bit from the fourth bit of a BCD (binary-coded decimal) number when BCD arithmetic is being used.

The P flag indicates the parity of the result in the accumulator, set for odd parity and reset for even parity — this is used mainly in connection with serial communications in which parity is being used. Finally, the C flag is used to indicate that a carry (addition) or a borrow (subtraction) in an arithmetic action is generated. All of these flag bits can be tested by software instructions, see Chapter 11. The flags have no direct effect on the operation of the microprocessor.

The other registers

Early MPUs had few registers other than the accumulator, but later designs have used large numbers of registers which have the same capabilities as the accumulator. These have been shown as one block in the diagram of Figure 12.1 and labelled as GP (general purpose) registers. Two registers, however, are of vital importance in any design of MPU, and these are labelled as *program counter* and *stack counter* (or *stack pointer*).

The *program counter* or *instruction pointer* register is used to hold address numbers. Initially, the PC/IP register will be set to the address of the first instruction that the MPU will execute, and the register is from then on incremented automatically so as to step from one instruction address to the next or to jump to a new address number and start another sequence from that point.

The PC/IP register is the means that the MPU uses to progress from one instruction to the next, and its automatic incrementing action forces the programming of the MPU to be done in steps. The two main exceptions to normal PC/IP action occur when the MPU is first switched on or reset, and when a jump occurs.

When the MPU is first switched on or reset, the chip design will force an address number to be placed into the PC/IP register. Older chips often used the number 0000H, but it is more common now to use a number at the top end of all the possible addresses. For a 32-bit MPU, this means the address FFFFFFF0H, for example. This means that the first byte of a program *must* be present at this address, and the usual instruction at this address is simply a *jump* instruction, forcing the PC/IP to use a different address where the initialisation program can be found. This is always an address in ROM.

The stack is used in conjunction with interrupts (see Chapter 9). When an interrupt signal arrives at one of the interrupt pins of the MPU, or is triggered by software, the first action that the MPU must carry out is to save the contents of all its registers. This is done by using a portion of the RAM that is designated as the 'stack' by placing its starting address in the stack register. The memory is used as first-in-last-out, and the terms push and pop (or pull) are used.

A push on to the stack will place the contents of a register into RAM at the address that is held in the stack register. Following this, the stack register number is decremented so that the next register will have its contents stored at the next available location lower in the memory, and so on until all of the registers have been saved. When the interrupt has been serviced, the register contents are returned, starting by incrementing the stack pointer number (because it will have been decremented to one step beyond the last memory address that was used), then recovering data, followed by several more steps of increment and load until all of the register contents have been retrieved and the stack pointer is at its initial address again. The use of the stack is dealt with in more detail in Chapter 11.

Finally, the block diagram shows a *multiplexer* stage. This is used when address data has to be sent along the internal data bus, and is applicable only if the internal data bus is narrower than the address bus.

Three-state bus (tri-state)

The MPU buses all feature three-state control. This does not mean a third logic state, only that all bus lines will be isolated, floating, when not in use. This allows the lines to be driven to whatever voltage they are connected to, rather than being kept at high or low logic level.

Reset circuits and pin

All MPU chips feature a RESET pin which allows the system to re-start in the event of a program failure. The same action can also be initiated by a software instruction. Many types of microcomputer use a reset button which will apply a voltage to the reset pin, but on recent models this has been discontinued on the grounds that a reset button could be pushed accidentally and would reset the machine with loss of all unsaved data, something that could be catastrophic for the user. On such machines, the RESET pin is used only in the start-up sequence. The action of the 8088 RESET pin is typical of most microprocessor chips. The reset action is started when the reset pin is forced high for more than four clock cycles — if the chip is being switched on initially, the reset pin must remain high for at least 50µs. Figure 12.5 shows a typical circuit connected to the RESET pin. The capacitor will ensure that the input voltage on the inverter will remain high

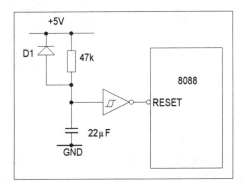

Figure 12.5 **The circuitry around the RESET pin of the 8088 processor**

until the capacitor charges to the voltage at which the inverter switches over. The diode will ensure that the capacitor is discharged when the machine is switched off.

The MPU will cease any actions when the reset pin is taken high and will remain inactive while the reset voltage is high. All buses are held in their floating state, and other pins are forced into preset logic voltages, low or high. When the reset pin voltage is taken low again, seven clock cycles are used for internal actions, and then the instruction pointer register is loaded with the fixed address FFFF0H, at which the first instruction must be located.

Support chips

Note: At this point you should revise the descriptions of parallel and serial port chips in Volume 2, Part 1, and also in this Volume.

PIO

The block diagram for a generalised PIO chip is illustrated in Figure 12.6. This is actually the block for one half of the Z80 PIO chip which in addition to these port blocks contains an interrupt control block and an internal control logic block.

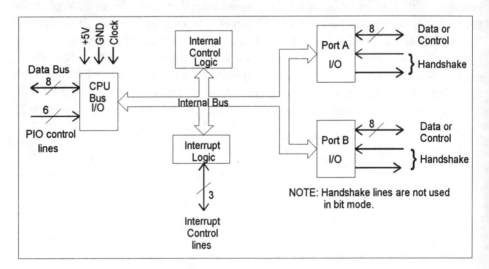

Figure 12.6   A block diagram for the internal arrangement of one port in the Z80 PIO chip

The 8-bit I/O bus at the right-hand side of the block diagram is the external bus which connects to the peripherals of the system. It is connected to two latching registers, the *data output register* and the *data input register*. Note that these are separate registers so that it would be possible for different bytes to be stored, one awaiting input to the microprocessor system, the other awaiting output.

The flow of data is controlled by an *input/output select register*. This select register is of eight bits, and controls the lines of the data input and output registers individually, rather than by using a register enable/disable control. This allows individual lines to be selected as inputs or outputs independently of the setting of other lines. This is particularly convenient if, for example, each input or output for a machine control system consists of, perhaps, just two bits. Each port can then be configured as four sets of two bits, which can be inputs or outputs.

The selection is controlled by a *mode-control register*, which selects the mode for a given port from the range of byte output, byte input, byte bi-directional (Port A only) or bit input/output. Since there are four possible choices here, the mode-control register needs to use two bits only. The remaining registers are used only when the bit input/output mode (mode 3) is selected.

On the microprocessor side of the ports, the chip connects to the Z80 data bus, along with three interrupt lines and six PIO control lines. Figure 12.7 shows a block diagram of the complete port chip.

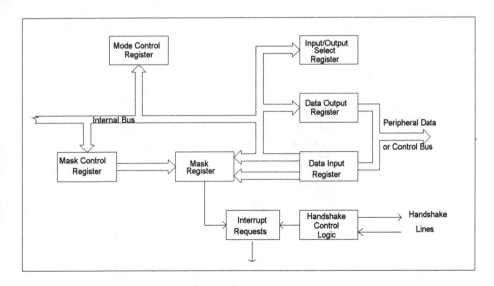

Figure 12.7 **The Z80 PIO chip block diagram — the port blocks represent the system illustrated in Figure 12.6**

Digital Techniques and Microprocessor Systems

Either of the user ports, labelled A and B, may be used for byte input, byte output or in bit mode with some output bits and some input bits on a given port. The A port also features a byte size bi-directional bus, not available on port B. The inputs and outputs are fully compatible with TTL logic levels, as they must be on any device of this type.

What makes the PIO so very much more complex is the feature of programmability. The Z80 PIO contains several registers which can be programmed from the microprocessor side of the ports, and which will thus dictate how the port will operate. This places much of the responsibility for the successful use of the port on to the shoulders of the software designer, leaving the hardware designer to concentrate on the connections to the interrupt lines, the correct control line connections, and the hardware that the port is intended to interface.

UART/ACIA

Serial input and output is handled by a UART or ACIA type of chip, and Figure 12.8 shows a typical block diagram. Only the internal data bus is illustrated in this diagram, the connections to the main MPU bus would be made through buffers under the control of the MPU.

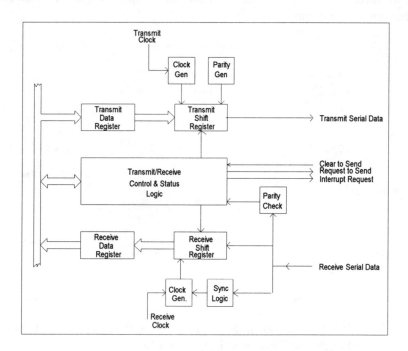

Figure 12.8   The standard UART/ACIA block diagram

228

The central portion of the UART/ACIA is the transmit/receive controller, which responds to CTS (clear to send) signals from the line, and will send out RTS (request to send) signals to the other devices. The signals are built up in shift registers. In the transmit shift register, a byte held in the transmit data register will be held and shifted out bit by bit at each pulse from the unit marked as clock gen. These units are often only dividers which reduce the pulse rate from an external clock (baud rate generator chip); the MPU clock is not used for this purpose.

When serial data is received, the complete byte is built up in the receive shift register, with one bit clocked in each time a pulse is received from the clock gen. unit — here again, this might be only a divider for an external clock pulse. When a byte has been assembled and parity (if used) checked, the byte is placed in the receive data register to be put on the data lines. The use of these data registers in addition to the shift registers is called *double-buffering*, and it avoids the need to read a byte rapidly from the shift registers in the time between the last bit of one byte and the first bit of the next byte.

Memory-oriented processors

MPU design was at one time classed either as memory-oriented or register-oriented, and diagrams of the register systems of two contrasting MPUs can be used to point out the differences in approach, though the vast majority of computing MPUs in use now are of the register-oriented variety. The following two diagrams show only register configurations, because the essential actions of the MPU are the same as for the general block diagram that has been illustrated earlier in this Chapter.

Figure 12.9 is a register diagram of the MCS 6502 MPU, a design which was very popular in the late 1970s and early 1980s. This type of processor was used in many early microcomputers by Commodore, Atari and Acorn, and was a predecessor of the modern RISC type of MPU, because it had a very small set of instructions, but each of these instructions could be executed very rapidly.

The main data registers of the 6502 are all of 8-bits. This is fairly normal for MPUs of this period, because 8-bit data was normal, but most other MPUs made provision for using some 16-bit registers, particularly for assembling address numbers. The 6502 assembled address numbers in two 8-bit actions and the index registers IX and IY were also 8-bit.

Index registers are used to hold numbers that are added to the address in the PC to form a new address (for a jump action), and using 8-bit registers in the 6502 limited the possibilities of a jump to 127 places forward or 128 places backward (because the number was interpreted as signed). Note that the status register is equivalent to the flag register of the example MPU at the start of this Chapter.

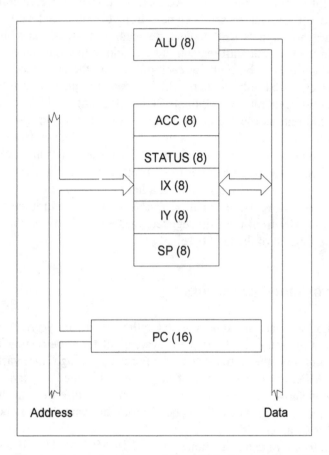

Figure 12.9 **The register structure of the 6502. It was possible to use this very small number of registers because the lowest 256 bytes of RAM (addresses 0000H to 00FFH) could be used as registers**

The PC is, of course, of 16 bits to allow addressing of RAM memory up to 64 Kbytes, but the use of the memory was restricted in ways that no other MPU has used since. The first 256 bytes of memory (Page 0 memory) were used, as noted above, as registers, with short and very fast commands capable of carrying out a limited set of actions on these registers. The second 256 bytes (Page 1 memory) were allocated for the stack, limiting the amount of stack that could be used and with no option for locating the stack elsewhere.

Register-oriented MPU — the Z80

By contrast with the 6502, the Z80, which appeared shortly after the 6502, follows the register-oriented pattern that has been used for most leading MPUs since, so that experience with the Z80 and its machine code programming is a good preparation for working with more recent MPUs. The Z80 also has the distinction of being the only MPU designed in the 1970s which is still used for some microcomputer designs in the 1990s.

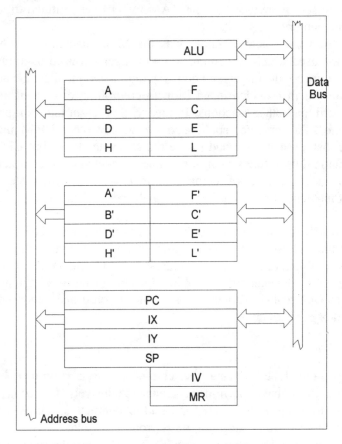

Figure 12.10   The Z80 registers, showing the system of pairing 8-bit registers so that they can be used as 16-bit registers

Figure 12.10 shows the registers of the Z80. The striking feature of this set is that all the important data registers exist in pairs, allowing you to work with 8-bit or 16-bit numbers. For example, the A register can be used as an 8-bit accumulator, as can the F register (normally used for flags), and the AF pair of registers can be used as a single 16-bit register. This allows immense flexibility to

231

the Z80 programmer, particularly since there are four pairs of these general-purpose registers, along with an alternate set (marked as A', F', and so on). These alternate registers are used by issuing a command which interchanges registers, allowing the A' to L' to be used in place of A to L or vice-versa.

The registers marked as PC, IX, IY and SP are all of 16-bit width. The IX and IY index registers will permit an index number of -37658 or $+37657$ to be added to the address in the PC register, so that longer-range jumps can be easily achieved as compared to the 6502. The stack pointer is also of 16-bits, allowing the stack to be set up anywhere in the RAM rather than confined to a specified portion of the memory.

The Z80, from manufacturer Zilog, was the MPU that, along with the Intel 8080, was first used widely in conjunction with a standardised operating system, CP/M. The letters are the initials of control, program and monitor, the actions that the operating system provided in addition to controlling disk input and output, video, keyboard, and other functions. In theory, a program that ran under CP/M control in one Z80 computer could also run in any other CP/M machine. In practice, the disk systems also had to use the same standards, but CP/M rapidly became a standard operating system for business-oriented microcomputing, and it ensured the quick demise of computers that were incompatible, with the exception of machines for which there was a closed market, such as education or games.

The IV and MR registers are the only simple 8-bit registers. The IV register is used to hold the address for interrupts (interrupt vector) so that each interrupt can be served by a different routine whose address is held in the IV register. The MR register is used as a counter for controlling dynamic memory refreshing, because the Z80 contains circuitry for memory refresh on board and needs no separate dynamic memory refresh chip.

Single-chip computers

Several chips aimed at reducing the chip-count of a computer have been produced, but these have been used almost exclusively for controllers and specialised instruments rather than for desktop computers. Three well-known examples are the Z8, from Zilog, the 6801 from Motorola, and the 80186 from Intel, along with several processors from Texas which are the most widely used of all controller chips.

Figure 12.11 illustrates the layout of the Z8. This is typical of this type of controller in that it contains both ROM and RAM, UART and ports, allowing it to be used in some applications with virtually no other chips. The RAM of 124 bytes can be used as registers in sets of 16, with the remainder used simply as RAM; another four bytes (making a total of 128 bytes) is used exclusively for ports. The ROM of the Z8 is masked, making it suitable for high-volume

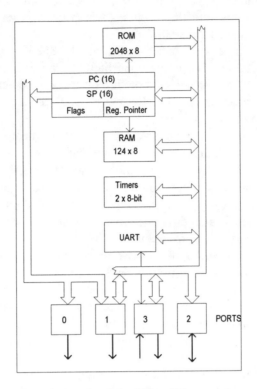

Figure 12.11   A block diagram of the Zilog Z8 processor unit

production, but there is a version designated Z86E21 which carries a socket piggy-backed on the chip so that EPROM can be used

The register pointer is used to select a set of 16 registers from the RAM, and any of such a set of 16 can be used as an accumulator or as an index register, as well as for general purposes. The stack can also be located in this 124 bytes of RAM. The program counter can address the internal ROM, or by using Ports 0 and 1 it can address external ROM (using Port 1 multiplexed between data and address). Port 3 can be used for serial or parallel connections with other units. External RAM can also be addressed, and control lines make it possible to use the same 16-bit address range for both ROM and RAM. The two timers are fully programmable and each uses a 6-bit counter.

Intel 8086

The Intel 8086 was one of the first 16-bit chips to be developed (in 1979), and at the time the difficulties in manufacturing it led to a 'cut-down' version, the 8088 being used in the IBM PC computer when it was introduced in 1981. The original

233

8086 version of the chip was used in later machines, and was the basis of the success of the first Amstrad PC machines which were faster and more capable than the designs that used the 8088.

Figure 12.12 illustrates the block diagram for the 8086 — note that this is not the form of the diagram used in the 8086 literature but has been redrawn in the style used for the 8088 to make the structure clearer. The diagram also shows the register structure which is considerably more complex that the 8-bit processors we have examined so far.

The 8086 design is dominated by the use of segmented addressing. Intel developed the 8086 with the intention of maintaining some compatibility with the older 8-bit chips such as the Intel 8080, around which the CP/M operating system was based. The 8086 is therefore designed to address memory in 64Kbyte portions, or *segments*, each of which is equivalent to the normal address space of an 8080 chip. This makes no difference to how memory is connected, only to the way that machine-code programs are written.

An address for the 8086 consists of two parts. One is a normal 16-bit address, the *offset address*, as would be used in an 8-bit processor; the other is a *segment address*, also of 16 bits. The segment address number is shifted four places left (one hex digit) and added to the offset address to form the 20-bit number that is placed on the address lines. Note that these lines are multiplexed, with the address and data sharing the lower 16 lines, and address and status bits sharing the higher 4 lines.

Though the use of segmented addressing slows the action of the processor, the compatibility that it brought to 16-bit computing allowed older software to be easily re-written for the 8086, and this greatly increased the amount of software available for the PC machine as compared to all others. This compatibility has been preserved, and the more recent chips from Intel have permitted a choice of segmented addressing, so that older programs can be run, and 'flat' addressing (using a 32-bit address number) that can be used with software using the Windows system.

Another feature of the 8086 is the use of an *instruction queue*. This is a set of six registers which feed the execution unit, and read instructions from memory in the time in a processor cycle when the address buses are not being used. This is a form of cache memory, and by ensuring that each instruction is in place in the queue it considerably speeds up execution times. This type of action has been used to a much greater extent in later Intel processors. Note that the 8088 used a four-instruction queue.

Data is fetched and operated on in word (two-byte) units, and this is also a feature that considerably increases execution speed as compared to older MPUs. This applies also to the 8088, which is 30% slower because of the use of 8-bit data units. Ports can be addressed separately from memory, using a 16-bit

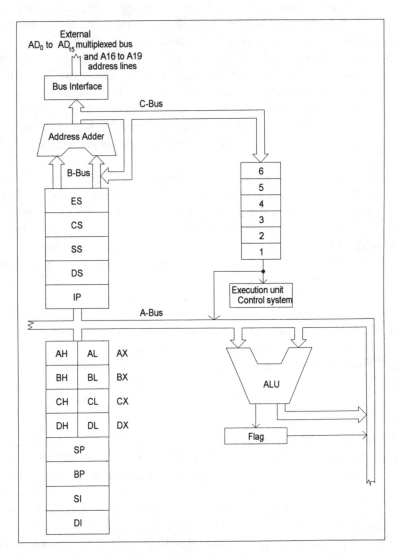

Figure 12.12   **A block diagram for the Intel 8086 chip, the first of a long line of compatible chips for the IBM PC type of computer. Only the registers that are accessible to a programmer are shown here**

address so that up to 65,536 ports can be addressed. Alternatively, memory addresses can be assigned to ports.

Clock pulses are generated by a separate chip, the 8284, and a co-processor, the 8087, can be added to speed up floating-point arithmetic by a factor of up to 100 where this is important (mainly in spreadsheet and CAD programs).

The register structure of the 8086 uses a bank of four 16-bit general purpose registers labelled as AX, BX, CX and DX, each of which can be used as if it were a pair of independent 8-bit registers such as AH and AL, using H and L to mean high bit and low bit respectively. The AX register is normally used as an accumulator, BX as a base address register for data addresses, CX as a counter, and DX for addressing ports. The register actions can, however, be interchanged if needed.

The flag/status register is also 16-bit, and registers BP, SP, SI and DI are index pointer registers holding address numbers. The SP register is the stack pointer. The IP register acts as a 16-bit program counter, holding the offset address, and the DS, CS, SS and ES registers are used for segment numbers. The use of four segment registers allows separate portions of memory to be used for different forms of data, so that you can use a data segment (DS register), a code segment (CS register), a stack segment (SS register) and an extra segment (ES register) if required.

Finally, the 8086 can be used in two modes. In minimum mode, the 8086 has control of the buses at all times, but in maximum mode, the bus control can be shared with other processors (usually the 8087 co-processor). The use of several pins is different in maximum mode, and a bus-controller chip, the 8288, must be used.

Intel 80486

The development of the Intel type of MPUs since the 8086 has been considerable, and the 8086 was followed by the 80286, 80386, 80486 and Pentium designs in that order. From the 80386 onwards, the programmer has been able to switch easily between segmented addressing and flat addressing using a 32-bit address bus, and the full-scale versions of the chips have all used a 32-bit data bus as well. The 80386SX MPU used a 16-bit data bus, and the 80486SX omitted the floating-point unit that was incorporated in the 80486DX. Pentium variations have concentrated on clock speed rather than on bus width or floating-point unit. All Pentium chips feature a floating-point unit (early versions suffered from a bug in this section), and use built-in cache memory for instructions.

Figure 12.13, at the end of this Chapter, shows the block diagram for the 80486DX to illustrate the complexity of this chip as compared to earlier designs. The diagram shows the complications that have been necessary to allow the dual addressing system, and also the much greater use of cache memory — the 80486 uses an 8Kbyte on-chip cache for address numbers so that the slower bus actions of fetching addresses from instruction words are carried out when the buses are otherwise not used.

In addition, the 80486 has been designed so as to allow for multi-program running, when several programs make time-shared use of the processor. This must provide for protection schemes which will prevent one program from corrupting the code or data of another. Since the 80486 was introduced, its life has been extended by the use of clock-doubling and clock-quadrupling chips that fit over the MPU. Using a 33MHz 80486 with the DX2 clock doubler allows the chip to run at 66MHz, and using a 25MHz 80486 with a DX4 quadrupler allows the chip to run at 100MHz. These designs have extended the life of the 80486 so that even at the time of writing it is competitive with its successor, the Pentium.

Test Questions

1. Which type of RAM, static or dynamic, would you expect to be used to implement a cache? State your reasons.
2. State two important differences between memory-oriented and register-oriented microprocessors.
3. Explain why most microprocessors specify a minimum clock rate, yet some can be operated down to very slow speeds.
4. What is meant by a microprogram? Name two advantages of using this system.
5. What is the essential difference between an instruction queue and a cache?
6. How are the ALU and the accumulator registers related?
7. How are the bits of the flag or status register used by the microprocessor?
8. Draw a circuit around a typical RESET pin for which the voltage must be held high for a minimum of 50µs after switch-on or switched reset.
9. Describe how a final address (effective address) is found for an 8086 processor when the segment address is A000H and the offset address is 1FC0H.
10. The 80486 chip can be switched to three different modes. In one mode, the address space is divided into 1Mbyte portions, and the code in one portion is not permitted to address any other portion. State why this is useful for compatibility with older processors.

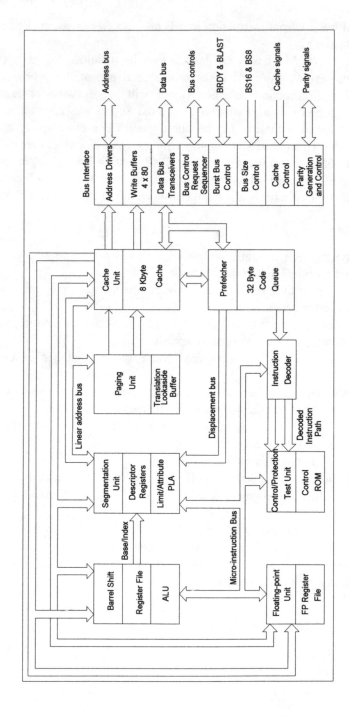

Figure 12.13 A block diagram of the Intel 80486 to illustrate the complexity of a modern microprocessor

13   Machine code programming

Syllabus reference, B1–C3, B2–D, M05

Machine code and assembly language

A microprocessor is controlled by a program which consists of a set of bytes of code numbers, called *machine code*. Each MPU type will use its own machine code, so that you cannot expect the code that operates one processor to operate another unless the MPUs are closely related like the Intel set. Some of these code numbers will be *operators*, the bytes that initiate an action. Other bytes will be *operands*, the bytes of data or bytes that provide the address of data for the operator. As far as you are concerned one byte looks like any other and the only way that an operator can be distinguished from an operand is in the order that they appear.

The correct order is therefore of the utmost importance. The first byte of a program must be a valid operator, and it must be followed by the correct number of operand bytes, and, in turn, by another operator and its operands. One omitted, misplaced, or incorrect byte makes a program useless; there is no 'almost right' as far as the order of program bytes is concerned.

Writing programs directly in number codes is simply too tedious for programs of more than a few bytes, and tedious actions are a certain cause of errors. For all but the simplest of machine code programs, therefore, a microprocessor program is written in what is called *assembly* (or *assembler*) language. This consists of abbreviations, called *mnemonics* (usually three letters each) for the commands, with each abbreviation representing a byte or several bytes of binary code. A program called an assembler will then convert these abbreviations into the number code for the microprocessor. Obviously, each different type of microprocessor needs a different assembler program.

The assembly language abbreviations for one type of microprocessor are reasonably similar to those of another, though the codes may be quite different. For example, the 6502 uses the mnemonic AND to indicate the instruction for ANDing the byte in the accumulator with a byte read from memory. The code (machine code, or object code) for this instruction, assuming that the next two bytes will be the full address of the byte in memory, is 00101101 (2DH). A similar instruction for the Z80 which assumes that the address in memory is held in a register pair, HL, is AND (HL), and its code is 10100110 (A6H).

It is therefore desirable as far as possible to work with assembly language mnemonics rather than with directly written machine code, so that the program can be more easily understood. In addition, the program using mnemonics (the source-code program) can be saved and edited, making it much easier to alter a machine code program by re-assembling the source code. An assembly language program can also be more easily converted into the form that another processor uses, using programs called *cross-assemblers*.

As far as possible, the elementary program examples in this book will be illustrated by examples using the 6502 and the Z80. These two designs, as has been remarked before, are radically different, and the ways in which they treat the use of memory are quite different. Anyone who has a smattering of the procedures for both of these very successful 8-bit microprocessors will have very little trouble in learning how to use any other microprocessor type, particularly the 16-bit micros which are based on the Z80 8-bit design.

Steps and bus actions

The operator and operand(s) of an instruction in assembly language correspond very closely to what happens on the MPU buses. Take for example, the write to memory action in Figure 13.1. We shall assume that the MPU is an 8-bit type (fewer bus steps would be needed for a 16- or 32-bit MPU). The assembly language indicates that the byte in the accumulator is to be written to a memory location which in this example is 7FF0H. The action for this particular MPU requires 11 distinct steps.

This set of steps is very important, and you should practice writing down the steps for load and save actions until you know them thoroughly. Note that these

```
                 Assembly language command: LD (7FF0),A

                      Machine code is 32F07F


  Step    Action                                  Comment
   1      PC          ────►  Address bus          address memory
   2      PC          ◄────  PC + 1               Increment PC
   3      Mem (32)    ────►  Data bus             Operator to instruction reg.
   4      PC          ────►  Address bus          Find next byte
   5      PC          ◄────  PC + 1               Increment PC
   6      Mem (F0)    ────►  Data bus             Operand low byte to memory pointer
   7      PC          ────►  Address bus          Find next byte
   8      PC          ◄────  PC + 1               Increment PC
   9      Mem (7F)    ────►  Data bus             Operand high byte to memory pointer
   10     MP          ────►  Address bus          Put operand number on address bus
   11     Data bus    ────►  Mem                  Save accumulator to address 7FF0
```

Figure 13.1   Bus actions for an assembly-language instruction

steps shown here are bus actions, they do not show the *internal* MPU actions such as connecting registers to buses. The number of steps will be different if the MPU can use a 16-bit or 32-bit data bus.

Instruction sets

The instruction set of a microprocessor is a list of the instructions for which it can be programmed, a sort of software specification. Instruction sets are invariably produced in the form of mnemonic*s*, accompanied by the instruction byte or bytes in hex code, of which the mnemonic is the more important part for the moment. Also needed is some explanation, however brief, of what the instruction does and what flags are affected.

For the experienced programmer, the instruction set represents a brief summary of how a microprocessor can be used, but for the novice it usually seems a huge and bewildering array of meaningless mnemonics with unfamiliar hex numbers. The hex numbers are not important if you are using an assembler program, and are there only to show what codes are being used.

Instructions are classed into groups according to the demands they make on memory and on registers. There is no agreement among manufacturers as to how these classifications are arrived at, but all instruction sets can be classed in similar ways. The main classes which can be used are:

1. *Load and Store* instructions. These form the largest group, and are concerned with transferring bytes between registers, or between register and memory. Port input/output instructions can be included in this group.
2. *Arithmetic and Logic* instructions. These include the usual add and subtract, AND, OR and X-OR instructions. In this group we can also include the instructions which increment or decrement a byte, and those which perform shift and rotate instructions that apply to registers or memory locations.
3. *Jump, Call and Return* instructions. These instructions act to cause a change in the address sequence of the program counter. A jump instruction is a one-way change; call and return allow a return to the normal program address sequence.
4. *Miscellaneous* instructions. These include the bit test instructions (is bit 5 set or reset?), HALT, NOP (no-operation), interrupt-enable or disable, and decimal correction instructions.

Load and Store

The set of instructions which are classed as LOAD and STORE include all of the instructions which feed data bytes into or out of the microprocessor. Most instruction sets use the terms LOAD and STORE, load meaning that data is to be read, and store meaning that data is to be written. Mnemonics such as LDA (load accumulator) and STA (store accumulator) are used in the 6502 instruction set, indicating that what is loaded is loaded into the accumulator and what is stored is the contents of the accumulator.

The Z80 instruction set uses only LD (load) for each type of transfer, and indicates the direction by the order of the mnemonics, so that, for example, LD A,01 means that the accumulator is loaded with the byte 01, and LD (223AH),A means that memory address 223AH is loaded from the accumulator. The round brackets are used to mean 'contents of'. The Intel processors use MOV in place of LD, but with the same order, so that MOV AH,02H means that the byte 02 is to be loaded into the AH register. The order of the operands is destination followed by source; you can think of the instruction as LOAD TO, FROM.

There are many variations that can be used for the load instruction. This is because:

(a) it may be possible to load into, or store from, a register other than the accumulator,
(b) when a byte is loaded from memory or stored to memory, there are always several methods by which the microprocessor can obtain the correct memory address (see later for addressing methods).

It is very convenient and useful to be able to load into, or store from, registers other than the accumulator. The 6502 allows the X and Y registers to be accessed

242

in this way. The Z80 has a very large number of instructions which permit any of its many registers to be loaded from memory or to memory. Equally, there is a large number of instructions for transfers between registers, all of which use the same LD mnemonic. The Intel chips use the MOV command for a huge number of data transfer actions among registers and memory.

Memory addressing methods

The main reason for the large number of instruction codes which appear in the instruction sets is the availability of many different methods for obtaining a memory address from which data can be fetched. These methods have evolved from computing experience, and it is very difficult for the learner to appreciate why some of the methods should be used, since the need for the less-obvious methods is likely to arise only when fairly advanced programming is carried out.

In this book we shall confine our attention to the methods which are most commonly used on the two example microprocessors, the 6502 and the Z80, using load and store instructions as examples. The Z80 addressing methods are used also on the Intel 8080 and on the Intel 8088, 8086, 80286, 80386, 80486 and Pentium family of processors. The examples that follow assume the use of an 8-bit processor that uses addressing of 16 bits.

One addressing method can be dismissed quickly. Implied addressing means that the address is implied in the opcode, and no other information is needed (so no operand is needed). Instructions which operate on one register only, or transfer or exchange between registers, are of this type, as are HALT and NOP (no-operation) codes.

Note that the full range of addressing methods is not necessarily available to each and every instruction in the instruction set.

Immediate addressing

The simplest addressing method which is used with a load instruction is immediate addressing. The number which is to be loaded into the register is located in the memory address which *immediately follows* the instruction code. The instruction therefore consists of a one-byte op-code and a one-byte operand, the operand in this case being the byte we want to load.

For example, if we wanted to load the number 1CH into the accumulator, and the load instruction were located at address 002AH, then the byte 1CH would have to be placed in memory address 002BH, immediately after the load instruction. The assembly language form of this instruction would be:

    6502   LDA #1CH
    Z80   LD A,1CH

There are no STORE immediate instructions. Note the different form of the mnemonics for the two different microprocessors. The 6502 uses the hashmark (#) to indicate immediate loading; the Z80 convention is that if the comma is followed by a single byte with no brackets, then immediate loading is to be used. The Z80 also permits the register pairs to be loaded with a byte each, as in examples such as:

LD HL,401AH

In this example, the H register is loaded with the byte 40H and the L register with the byte 1A; the letters H and L are used to remind you of High byte and Low byte.

Immediate loading is rapid, because the memory address for the data is obtained simply by incrementing the program counter. It is not necessarily a convenient method, however, because the numbers which are to be loaded in by this method must be part of the program itself so that they will be placed in the program memory at the time when the program is entered into memory. Immediate loading is useful when a program must contain constants (as when a byte in the accumulator is to be compared with some constant). If the program is in ROM and the data is all in RAM, this type of addressing cannot be used.

Absolute, extended, or direct addressing

All other forms of memory addressing are designed to address memory which is located at an address which is remote from the address of the instruction. Of these other methods, absolute addressing (also called extended or direct addressing) is the most straightforward.

An absolute load or store instruction consists of three bytes — an op-code byte and two operand bytes. The first byte (some instructions may require two bytes) is the instruction, which will specify by its coding which register is to be used, and this is followed by two bytes which specify the address at which the data is to be found or placed.

This straightforward scheme is slightly complicated by a convention used by all microprocessors in which the address is written lower-byte first. For example, the address 043FH would be placed in program memory as 3F04. This is a worry only when programming is done by hand, because an assembler will carry out the reversal automatically. To load from an address 062CH, the mnemonics would be:

6502   LDA 062CH which in machine code is AD2C06

Z80   LD A,(062CH) which in machine code is 3A2C06

Note that the Z80 assembler convention uses round brackets to mean 'contents of' and the order A, (memory address) to indicate the direction of transfer. The corresponding store instructions are:

6502   STA 062CH generating the code 8D2C06

Z80   LD (062CH),A generating the code 322C06

The Z80 uses the same mnemonic LD for both loading and storing, but the reversal of the order implies that the contents of the accumulator are transferred to the memory address. The Intel chips use the MOV instruction in the same way.

Absolute addressing allows a byte to be loaded from or stored to any memory address which consists of two bytes (or as many bytes as are required — modern processors will require a 32-bit address). The method is, however, time and memory consuming. Three bytes of program code are used for 8-bit MPUs, taking up three bytes of program memory, quite apart from the memory used for the data, and the time needed to carry out the instruction can range from four clock cycles (6502) to seven clock cycles (Z80). In addition, the address which is to be used has to be known at the time when the program is written — you cannot alter it without altering the program.

Zero-page addressing

A modification of absolute addressing which is used on the 6502 and to a much lesser extent on the Z80 is 'zero-page addressing'. The name zero page refers to the upper byte of the address number, which is zero for a zero-page address. The range of addresses covered by this method is therefore 0000H to 00FFH, a total of only 256 (denary) addresses. A 6502 zero-page address instruction needs only two bytes of program memory and takes three clock cycles to execute.

For example, the assembler instruction LDA 3FH to a 6502 assembler will instruct the microprocessor to load the accumulator from memory address 003FH. The Z80 uses zero-page addressing only for a set of RST (reset) instructions, but the 6502 has zero-page addressing available for most of its instruction set commands. Programmers who use the 6502 therefore keep the most-used bytes (as far as possible) in the memory which has zero-page addresses, using these addresses as if they were registers.

Indirect addressing

Indirect addressing is a method which can be used to great advantage by an experienced programmer, because the instruction does not have to specify the memory address, simply where numbers that make up the address can be found. The form of indirect addressing which is available (and much-used) on the Z80 is register-indirect addressing, and can make use of any of the register pairs, BC, HL or DE, though the HL pair is used more frequently than the other two, because more commands are available that use these registers.

The command LD A,(HL) means load the accumulator from the memory address which is contained in the HL pair of registers (the high byte in H, the low byte in L). Similarly, LD A,(BC) and LD A,(DE) mean that the accumulator is to be loaded from the address contained in the BC or DE register pairs respectively; other registers (other than the accumulator, that is) can be loaded only from the address contained in the HL pair. The corresponding store instructions are for the LD(HL),A; LD(BC),A; LD(DE),A. Since these registers can be loaded from memory, this provides a way of using an address that can be changed, since it is part of the data rather than part of the program.

Each of these indirect instructions needs seven clock cycles, but requires only a single-byte instruction in which the actual memory address need not be specified. It is up to the programmer, however, to ensure in preceding program steps that a suitable address is placed into the HL register pair before the load instruction occurs. The 6502 does not have any register-indirect memory addressing of this type.

Indexed addressing

Indexed addressing is a feature of most microprocessor instruction sets but tends to mean different things to different designers. As used on the 6502, indexed addressing can be absolute or indirect, and the index registers are 8-bit only. In an absolute indexed address instruction, the instruction code is followed by two bytes which specify an address.

The actual address which is used, however, is obtained by adding the content of an index register (X or Y) to the specified address. Suppose, for example, that the X register of the 6502 contained the byte 0AH. The instruction

LDA 326FH,X

will cause the accumulator to be loaded, not from address 326FH but from 326FH + 0AH, which is 3279H. The instruction needs three bytes and four clock cycles; its usefulness lies in the fact that a range of different addresses around a central address can be obtained by using the same instruction with a different number in the X-register. This is particularly useful when an instruction for loading forms part of a loop, and the address number has to be incremented on each pass through the loop, or when the byte which is fetched must correspond to the byte in the X-register.

For example, a code translation system becomes easy if the input bytes of code are numbers which, when placed in the X-register, fetch the correct codes from the address. A brief example is illustrated in Table 13.1 overleaf. The illustrations of indexing have used the X-register in this instance but the 6502 can also use the Y-index register.

The 6502 can use two forms of indirect addressing along with the two index registers, X and Y. *Indexed indirect* addressing makes use of the index register X, and addresses in the zero page of memory (0000H to 00FFH). In an indexed indirect instruction, a byte is taken from page zero memory, at an address which is specified by the operand, the second byte of the instruction. This byte address is specified by adding the operand to the contents of the X-register.

Table 13.1   Illustrating the use of indexed addressing when a translation is needed — in this case using small numbers to refer to ASCII codes

INDEX X	RAM Address	Content of RAM	ASCII letter
05H	0405H	4DH	M
06H	0406H	57H	W
07H	0407H	44H	D
08H	0408H	5AH	Z
09H	0409H	41H	A
0AH	040AH	54H	T
0BH	040BH	49H	I
...	...	...	...
etc.	etc.	etc.	etc.

Having obtained this zero-page address, the byte at this address constitutes the low order byte of an address, ADL. The zero-page address is then incremented, to load in another byte which will be used as the high order address, ADH.

For example, if the X register contains 05H, then the command LDA (15H,X) causes the number 0015H (a zero-page address) to be added to 05H (the number in the X-register) to give 001A.This is put on to the address and fetches a byte, for the sake of example, 26H. The address lines now increment to 601BH, and another byte is fetched, we shall imagine that it is 3AH. These bytes are now placed on the address lines to give the address 3A26H (remember that the number has been stored in low–high order), which is the address from which the data will be fetched or to which it is stored.

The other form of 6502 indirect addressing is called *indirect indexed* (note the reversal of the words), and uses the Y-register. The instruction consists of two bytes. The second (operand) is, as before, a zero-page address. The address is put on to the address lines, and the byte fetched from this address has the contents of the Y-register added to it, and is then stored as the address low byte. The address high byte is then taken from the next consecutive address in zero-page memory. Finally the address bytes are placed on the address lines so that the data can be loaded or stored.

As an example, imagine that the Y-register has been loaded with 0AH. The instruction is LDA(15),Y, and the first part of the action is to put the address 0015H on to the address lines. This might, for example, fetch a byte 14H. The contents of the Y-register are added to this to give 1EH, and the zero-page address is incremented to 0016H. If this address fetches in the byte 2AH, then the address which is put on to the lines now is 2A1EH, and it is this address which will be read from or written to finally.

Indexed addressing on the Z80 is much more typical of the majority of other processors, including the modern types, and makes use of the two 16-bit index registers, IX and IY. An indexed load instruction must include the name of the register and also a displacement byte which will be added to the address in the index register. For example, the instruction

$$\text{LD A,(IX + 0AH)}$$

will cause the accumulator to be loaded from the address in the IX register plus 0AH. If the address in the IX register is 127AH, then the LD A,(IX + 0AH) instruction will load the accumulator from 127AH + 0AH = 1284H.

Indexed addressing is used mainly on the Z80, as on the 6502, when a series of addresses has to be accessed. If, for example, memory location 217BH contains the first byte of a set of numbers which are to be loaded in succession, then by placing this address in an index (using IX in the Z80, and a pair of zero-page address numbers in the 6502) successive bytes can be loaded by using the same instruction, with the displacement increased by one (incremented) each time the instruction is executed. This is particularly suited to loop programming in which a set of instructions is repeated. For example, if a set of successive memory addresses stores the ASCII codes for 'That's all, folks', a loop instruction can load in byte by byte and send each byte to the video system to be displayed on the screen.

One complication relating to indexing is the use of negative numbers. Displacements of up to 7FH are treated as positive, displacements of 80H to FFH are taken as negative, following the usual convention of a 2s complement number.

Relative addressing

Relative addressing is a method which is used for only a few 6502 or Z80 instructions. A relative address is fixed by adding a displacement byte (00H to FFH) to the address which is stored in the program counter at the time of the instruction. This type of addressing allows a change of address as large as can be obtained with the displacement byte, and for signed single bytes this can be from −128 to +127 steps of address number. 16-bit processors can use 16-bit displacements, and in the Intel range this is complicated by the possibility of moving from one segment to another. The more recent 32-bit processors can use relative addressing with 32-bit displacements. Relative addressing will be looked at in more detail in connection with jumps, see later.

Register-to-register transfers and exchanges

The load and store instructions all require some reference to memory, but transfers or exchanges between registers do not, and are therefore single-byte instructions which can be executed rapidly. The difference between a transfer and an exchange is that in a transfer, the contents of one register are replaced. For example, if a microprocessor has 8-bit registers A and B, and the byte in A is 0AH, the byte in B being 1FH, then the effect of an exchange instruction will be to place 1FH in A and 0AH in B. The instruction that will transfer contents of B to A will result in both registers containing the byte 1FH.

Table 13.2   The register transfer instructions of the 6502

Mnemonic for assembler	Action
PHA	Push A to stack
PHP	Push Status/Flag register to stack
PLA	Pull from stack to A
PLP	Pull from stack to Status/Flag
TAX	Transfer A to X
TAY	Transfer A to Y
TXA	Transfer X to A
TXS	Transfer X to SP
TYA	Transfer Y to A

The transfer instructions of the 6502 are shown in Table 13.2. The registers which can be used are the accumulator A, the index registers X and Y and the stack pointer S (this is how the stack pointer can be changed). There are no register exchange instructions. By contrast, Table 13.3 shows the transfer and exchange set for the Z80. The contents of any register can be copied to any other register, and a few exchange instructions exist, notably the exchange of the register pairs DE and HL.

The Z80 instructions EX AF,A'F' and EXX have no counterpart in any other commonly-used microprocessor; they allow the second alternate set of registers to be used in place of the normal set. You cannot use the normal set and the alternate set together, however. In addition, a set of block transfer and search instructions enable one command to be used for a routine which in any other 8-bit microprocessor would require several commands in a loop program.

Table 13.3   The transfer and exchange instructions of the Z80. The register 'R' can be any register except the I register

Mnemonic	Action	Mnemonic	Action
LA A,R	Load A from any register R	LD SP,HL	Load Stack pointer from HL
LD B,R	Load B from any register R	LD SP,IX	Load Stack pointer from IX
LD C,R	Load C from any register R	LD SP,IY	Load Stack pointer from IY
LD D,R	Load D from any register R	EX AF,A'F'	Exchange AF with alternate AF
LD E,R	Load E from any register R	EX DE,HL	Exchange DE and HL register contents
LD H,R	Load H from any register R	EXX	Exchange DE,HL,BC with alternate set
LD L,R	Load L from any register R		

Logical and arithmetical instructions

These instructions are the main 'doing', as distinct from shifting, instructions of the microprocessor, and their actions are fairly well standardised. Taking the logical instructions, AND, OR, XOR first, each of these carries out its logic function *bitwise*. That means that each bit of a byte taken from memory or from another register is gated with the corresponding bit of the byte in the accumulator, and the result of the gating action is stored back in the accumulator. The AND function, for example, follows the rule 1 AND 1 = 1, with any other gated pair producing 0. For example, when the byte 3BH is ANDed with the byte 6FH, the result is 2BH. The detailed action is shown in Figure 13.2, with the bytes written out in binary form and arranged in columns.

This action, like all the actions in this group, can affect some of the flags of the status register, in particular the zero flag. The 6502 will set the N-flag if there is a 1 bit in the highest place of the accumulator; on the Z80 only the Z and S flags are affected. Note that AND, OR and XOR actions are *not* additions, they cannot produce a carry and cannot affect the C flag.

Apart from the use of these instructions in the logic of machine control, programmers make use of them to short-circuit procedures. For example, the Z80 command XOR A will clear the accumulator, and takes less time and program space than LD A,0. The Z80, unlike the 6502, does not change flags in the status register when a byte is loaded to the accumulator, so that to test if a byte is zero or negative after a load, the instruction OR A can be used. This has no effect on the byte itself, but does allow the flags to be set or reset according to the nature of the byte in the accumulator.

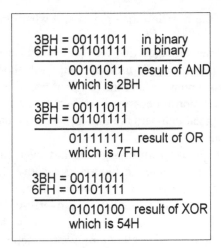

Figure 13.2   The results of AND, OR and XOR actions in binary and hex

Other dodges of this type include AND 0 (clears the accumulator), AND FFH (leaves the accumulator unchanged, sets flags) and OR 0FFH (fills the accumulator with ls.). The reason for using these methods is that they are faster than the more conventional alternatives.

Arithmetic commands

The arithmetic commands consist of addition and subtraction only for the majority of 8-bit microprocessors. Most 8-bit micros also permit only 8-bit arithmetic, which is performed in the accumulator, but the Z80 permits a few 16-bit operations using the register pairs HL with BC or DE.

The arithmetic instructions will affect the carry and overflow flags as well as the zero and negative (sign) flags, and provisions have to be made for using the carry as well. When two 8-bit numbers are added, the result may be a 9-bit number, which will cause the carry bit to be set. This carry bit can then be added into the next pair of numbers which are added, so that the addition of a 16-bit number is automatically correct, providing that the lower order bytes are added, then the higher order bytes (the reason, incidentally, for the low–high order).

Overflow

Overflow is a concept which is important only if arithmetical programs are being designed from scratch. In general, programs for all standard operations are standardised and obtainable from the manufacturers of the microprocessor, so that the programmer will not normally have to 're-invent the wheel'. It is useful to

251

know, however, why the overflow flag exists and what circumstances cause it to be set or reset.

When the programmer assumes that all the bytes in arithmetic processes are signed, that is, the highest order is used as a sign bit, then only 7 bits of a byte can be used to express a number, and the range of possible numbers is from −128 to +127 (denary). The microprocessor, however, treats all numbers in the same way, and when two signed numbers are added, the usual carry from one bit place to the next will be done.

If the addition causes a carry to the sign bit, however, the sign bit of the result may be incorrect, as the examples in Figure 13.3 show. This is indicated to the programmer by setting the overflow bit, so that a section of program can be written to detect this error and correct it. The overflow bit does not cause any automatic action, unlike the carry bit, it is purely for the programmer's convenience and will be ignored in all but a few arithmetic programs.

```
(a)      01111111      127 denary
         00001111      15 denary

         10001110      -114 (signed)


(b)      10010101      -107 denary
         10001101      -115 denary

carry 1  00100010      +34 denary

The OV flag is set by the XOR of carries
```

Figure 13.3   **Examples which set the OV bit as a warning. In (a) there has been a carry from the 7th bit that alters the sign bit. In (b) there has been a carry from the 8th bit which sets the carry flag and also changes the sign bit**

Shifts and rotations

Shifts form another useful microprocessor action, and all types of micro-processors incorporate shift and rotate instructions, but these vary considerably in number and complexity from one design to another.

The 6502 uses only two shifts, LSR and ASL. The *LSR* (logical shift right) instruction moves each bit in the accumulator or in a specified memory address one bit to the right (high to low place), with a zero shifted into the highest order place, and the lowest bit shifted to the carry. Several of the range of 6502 address methods are available if the byte is not in the accumulator. The *ASL* (arithmetic

252

shift left) causes a left shift, with a zero loaded into the lowest order bit, and the highest order bit shifted to the carry. Once again, several addressing methods are available if the byte is not in the accumulator.

The rotate instructions of the 6502 are ROL and ROR. Each of these uses the carry flag as part of the rotation, and the byte which is rotated can be in the accumulator or in a memory address. The *ROL* instruction rotates left with the highest order bit transferred to the carry, and the carry bit transferred in turn to the lowest order of the accumulator or memory byte. The *ROR* instruction rotates right, with the lowest order bit transferred to the carry, and the carry bit transferred to the highest order place of the byte.

The Z80 offers a large variety of shift and rotate instructions, permitting left or right shift or rotation, including or excluding the carry and with one curious instruction (SRA) which leaves the highest order bit unshifted. The Z80 shift and rotate instructions are summarised in Table 13.4.

Table 13.4 The large set of rotate and shift actions for the Z80

Mnemonic	Action
RLCA	Rotate accumulator byte left, bit 7 copied to carry and also to bit 0
RLA	Rotate accumulator byte left, bit 7 to carry, carry to bit 0
RRCA	Rotate accumulator bit right, bit 0 copied to carry and also to bit 7
RRA	Rotate accumulator bit right, carry to bit 7, bit 0 to carry
RLC R	Rotate left byte in register R, bit 7 to carry and to bit 0.
RLC(HL)	Rotate left byte addressed by HL pair, bit 7 to carry and to bit 0
RLC(IX+d)	Rotate left bit addressed by index, bit 7 to carry and to bit 0
RLC(IY+d)	Rotate left bit addressed by index, bit 7 to carry and to bit 0
RL R	Rotate left byte in register or addressed by register pair. Bit 7 to carry, carry to bit 1
RRC R	Rotate right byte in register or addressed by register pair. Bit 0 to bit 7 and also to carry
RR R	Rotate right byte in register or addressed by register pair. Bit 0 to carry, carry to bit 7
SLA R	Shift left byte in register or addressed by register pair. Bit 7 to carry, 0 to bit 0.
SRA R	Shift right byte in register or addressed by register pair. Bit 7 unchanged and copied to bit 6, bit 0 to carry.
SRL R	Shift left byte in register or addressed by register pair. Bit 0 to carry, 0 into bit 0
RLD	Acts on byte addressed by HL. Lower nibble copied into upper nibble, upper nibble to accumulator low, accumulator low into lower nibble of byte addressed by HL, accumulator high unchanged.
RRD	Acts on byte addressed by HL. Lower nibble of byte copied into accumulator low, accumulator low into higher nibble of byte, higher nibble of byte to lower nibble of byte. Accumulator high nibble unchanged.

Using the add and subtract commands, along with rotation and shifting, any mathematical process can be performed on binary numbers. The routines which are needed are often very complex, but since they are all well-established, they can be included into programs when needed.

Flowchart and pseudo-language

Planning is a very important part of any assembly-language program, and the traditional method of planning is the use of a flowchart for each portion of a program. The more modern method is called pseudo-language or pseudo-code. Both are illustrated in Figure 13.4.

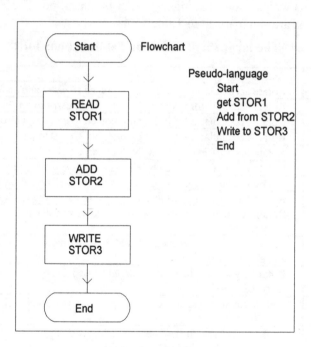

Figure 13.4   The flowchart and pseudo-language for a load, add, and save action

Flowcharts use a set of standard symbols, some of the most important of which are illustrated in Figure 13.5. The use of flowcharts is well established, because they have been used since the beginning of computing, but they are now regarded as old-fashioned, and pseudo-language is much more common. The advantages of pseudo-language are that the words and the steps are much easier to translate into terms of assembler language (or higher level languages), and that less space is

254

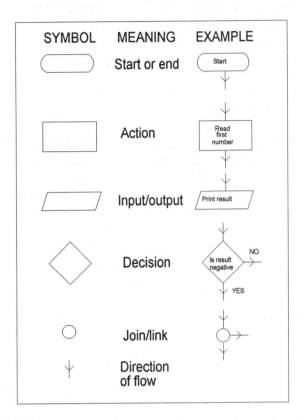

Figure 13.5   The main flowchart symbols

needed. In addition, pseudo-language is less formal, so that it can include notes (such as where a quantity will be stored).

The planning of a program should start with the main steps, whether you use flowchart or pseudo-language methods, and then each main step should in turn be treated as if it were a program on its own. This system of breaking down the program into steps should continue until the steps are small enough to allow you to write assembly language for each step. The general rule of thumb is that a step is small enough when the assembly language occupies only one page or less of text.

When a program is broken down in this way, it is likely that some of the steps will be identical to ones that you have used before, so that you can re-use assembly language, perhaps with changes to address numbers. This is important, because tried and tested code is always preferable to new and untested code. In addition, this encourages the use of sub-routines, see later.

255

NOTE: In the examples of assembly language program steps that follow, you should write plans both in flowchart form and in pseudo-language for each. Avoid being too specific about flowchart steps — this error is often referred to as writing 'code in boxes'.

Looping and branching

So far, the program operations which we have looked at have included loading and storing, interchange, arithmetic, and logical operations. Writing a program for a microprocessor, assuming that the program has been fully planned, then consists of writing down the assembler mnemonics for the sequence of operations that will be required. Two simple examples (not of the same program) for both Z80 and 6502 are shown here.

```
Z80                    6502
    LD  HL,3802H           LDA  1744H
    LD  A,(HL)             AND  #1F
    OR  01H                ORA  #01H
    LD  (HL),A             STA  00EEH
```

These simple programs have one factor in common — they start, carry out a sequence of instructions, then stop, having progressed from a low address number to a higher one. This type of program is called *linear*, and a number of programs are of this type, but much more useful and interesting effects can be achieved if a program is capable of branching (jumping from one address to another which is not in sequence) and looping (repeating a series of program steps several times).

Imagine, for example, one portion of what is needed in a program to activate a set of traffic lights. The red light has just switched on, and it must remain on for a set period before the amber light is also activated. The set period of delay is obtained by loading a number into memory or into a register, checking that it is not zero, decrementing it, and then checking again.

In program form, this process is shown in Figure 13.6, using comments to indicate what is happening. NOTE the use of a semicolon to indicate a comment. The assembler will ignore anything that follows a semicolon in an instruction line.

Since the speed of each part of the operation is controlled by the clock rate, the timing will be as precise as the clock itself; since the clock is usually crystal-controlled, the timing can be very precise indeed. Each time the checking operation is carried out and the number is found to be greater than zero, the program 'loops back', meaning that the program counter is returned to the address of the operation which carried out the decrementing of the number.

```
Z80                                           6502
                                                       CLD            ;clear decimal mode
        LD A,0FFH       ;FF into A                     LDX #0FFH      ;load in FFH  to X
AGAIN:  DEC A           ;decrement             AGAIN:  DEX            ; decrement X
        JR NZ, AGAIN    ;back if not zero              BNE AGAIN      ; back if not zero
        HALT            ;stop                          BRK            ;stop
```

Figure 13.6   A delay routine using a loop. The program branches back to the address labelled as AGAIN for as long as the register content is not zero after the decrement step

This simple routine also illustrates a *branch* or *jump*. Each time the number is decremented and found to be greater than zero, the program loops back, but when the number is zero, the program continues, allowing the program counter to reach the address beyond the branch instruction. In practical terms, this is the step which will switch the amber light on; in program terms, the program puts a new address on the lines so as to pick up a new instruction.

Note that the assembler can use a *label* name to indicate a program address number. Where the word 'AGAIN:' appears, the assembler takes the colon sign to mean that the word ahead of the colon is a label name which means the address number of the instruction that follows. When the word AGAIN is used later (without the colon) as if it were an address, the assembler will insert the correct address into the code.

The ability to branch to a new address depending on a test of a number in a memory (particularly 6502) or in a register (particularly Z80) is such an important feature of the use of the microprocessor that we shall devote considerably more space to it. The tests which all microprocessors can carry out are for the tested quantity being zero or non-zero, positive or negative, carry bit set or reset. This list might not seem promising, but these few comparisons make possible the large range of microprocessor decision steps, and each decision can be made in a number of ways, according to the intentions of the programmer. Later types of microprocessor can carry out a larger range of tests.

For example, still using the principle of a variable number in the accumulator which is being decremented, the test could be such that the program looped back for as long as the number in the accumulator was positive, or until it became zero, or that it did not loop back when the number became negative. The decision regarding which method to use is one based purely on convenience; it often happens that one method will result in a shorter or less clumsy program than another, and is therefore preferred.

Comparison steps

We've said that the simple comparison steps of a microprocessor are sufficient for all decision processes, and this requires some explanation. Suppose that, at the decision step, the microprocessor has to branch to one address if the number in the accumulator is 0BH, but not if any other number is present. This type of comparison needs the use of an additional instruction, CMP or CP, which is used by both the 6502 and the Z80. Figure 13.7 shows an example of how this comparison might be done, using the immediate form of the CMP or CP instruction (CMP 0BH).

```
Z80                                        6502
    LD A,(HL)     ;byte loaded                 LDA 0050H     ;load from address 50H
    CP 08H        ;is it 8?                    CMP 08H       ;is it 08?
    JR Z,USEIT    ;jump if it is               BEQ USEIT     ;jump if it is
    LD A,(DE)     ;try another byte            LDA 0060H     ;try another byte
```

Figure 13.7   Using a comparison step for recognition. This instruction sets flags but does not alter the contents of registers

The action of the CP or CMP instruction is to subtract the byte, in this example 08H, from the byte in the accumulator, and set the status flags accordingly, but unlike a subtract instruction it does not alter the byte in the accumulator. A subtract instruction used at this point would leave the accumulator filled with the result of the subtraction. Since the flags are set by the CP/CMP instruction, then the BRANCH-ON-ZERO instruction can be used following CMP to ensure that the program will branch to a new address if the byte in the accumulator has been 08H.

For any other number, the result will be either positive or negative but not zero, so that no branch can take place. Because the CMP instruction leaves the contents of the accumulator unaffected, the program can then continue normally, and the byte in the accumulator can be used if needed.

The byte which is compared may, of course, be the ASCII code for a letter. If a portion of program such as is illustrated in Figure 13.8 is used, assuming a monitor program which will cause the ASCII value of a depressed key to load into the accumulator, then a looping routine can be forced to branch only when the correct letter has been entered.

This can be extended by using more branch steps so that a complete word must be entered before the final branch (Figure 13.9) and this is one of the techniques which is used to write a program which converts word inputs into micro-processor instructions — the essential part of a computer language program.

```
Z80                                         6502

GETONE:  LD A, KEYPRES    ;get byte         GETONE:  LDA KEYPRES    ;get byte
         CP 49H           ;is it 'I'?                CMP 49H        ;is it 'I'?
         JR NZ, GETONE    ;back if not               BNE GETONE    ;back if not
```

Figure 13.8   Finding when a particular letter key is pressed

```
Z80                                    6502

GET1:  LD A; KPRES                     GET1:  LDA KPRES
       CP 41H          ;letter 'A'            CMP #41H        ;letter 'A'
       JR NZ, GET1                            BNE GET1
GET2:  LD A, KPRES                     GET2:  LDA KPRES
       CP 43H          ;letter 'C'            CMP #43H        ;letter 'C'
       JR NZ, GET2                            BNE GET2
GET3:  LD A, KPRES                     GET3:  LDA KEYPRES
       CP 54H          ;letter 'T'            CMP #54H        ;letter 'T'
       JR NZ, GET3                            BNE GET3
       ;Rest of program follows               ;Rest of program follows
```

Figure 13.9   Using the recognition loop to find a complete word. This is a primitive example and there are better methods, using a loop, which are beyond the scope of this book

The example in Figure 13.9 is *very* primitive, and in a real-life program, a loop would be used to fetch and test the bytes that make up the word. The point we need to grasp, however, is that each character of a word must be tested — there is no way of recognising a complete word.

Unconditional jumps

The BRANCH commands which depend on the setting of flags are called conditional branches or conditional jumps, for obvious reasons. Microprocessors also permit unconditional jumps, which cause the address to change irrespective of any other considerations. Unconditional jumps are used to prevent one section of program automatically progressing to another.

For example, it is often necessary to place a short section of program (for reasons which will be explained later) between two existing sections, but without interrupting the normal flow of the program. This can be done by using an unconditional jump instruction at the end of the first 'broken' section which will transfer operations to the start of the second section, bridging the inserted section of steps (Figure 13.10). The inserted section can then only be accessed by a jump to its starting address, and can be arranged so that it will jump to another address on completion.

```
         Z80                                        6502
              LD A,(HL)                                  LDA SOURCE
              JR OVER         ;jump next part            JMP OVER         ;jump next part
     ISLAND:  LD A,(DE)                         ISLAND:  LDA OTHER
              CP 20H                                     CMP #20H
              JR Z,OUT                                   BNE OUT
     OVER:    LD(NEXT),A      ;jump to here     OVER:    LD NEXT          ;jump to here
```

Figure 13.10   Using an unconditional jump to bridge a section of code

This insertion technique is easy only when an assembler is used for programming — when hand-programming is used, the program must usually be re-entered into memory from the break point onwards so that the addresses are correct.

The two microprocessors which have been used to illustrate techniques both treat jumps in a fairly similar way. There are differences between the two microprocessor types, however, in the treatment of conditional jumps. The Z80 has a set of conditional jump instructions which are coded on the assembler as JP followed by the condition and an absolute address. For example, the instruction:

JP NZ,036H

means 'jump if the content of the accumulator is not zero, to the address 0036H, otherwise continue'. The 6502 has no instructions of this type, since all conditional jumps (called branches to distinguish them) are program-relative, using a displacement byte to indicate the number of steps on or back that must be jumped.

Timing loops

A rudimentary timing loop was illustrated in Figure 13.6. The principle is simple: a number is loaded into memory or into a register, decremented and tested, with a loop back to the decrement step until the number is decremented to zero. The total time needed can be found by adding up the times for the instructions within the loop, and then multiplying by the number of times that the loop is used, see Figure 13.11.

Z80		6502	
	LD A,FF		LDX #FF
BACK:	DEC A	BACK:	DEX
	JR NZ,BACK		BNE BACK

DEC	23 cycles	DEX	2 cycles
JR NZ	7 cycles if	BNE	3 cycles
	condition not met		
	12 if met		

Since FFH = 255, the condition is not met 254 times, so that the time is 23 x 255 for DEC and 7 x 254 + 12 for JR NZ, making a total of 7,665 clock cycles.

Total time is (3+2) x 255 = 1,275 clock cycles.
NOTE that this illustrates the speed of a processor with a small fast set of instructions

Figure 13.11   Calculating the total time for a loop, and also illustrating the RISC principles

This time can be extended by using more instructions within the loop or by using another timing loop within the original loop (or enclosing the original loop inside another). Figure 13.12 illustrates two methods — on the Z80 the double register can be used to provide a larger number to count down, and on the 6502, two registers are loaded, and one is counted down completely before the other is decremented and tested. Obviously you can use a combination of both methods on the Z80 and later processor types.

```
Z80                          6502
        LD BC,FFFFH                  LDY #FFH
BACK:   DEC BC               LOAD:   LDX #FFH
        LD A,B               LOOP:   DEX
        OR C                         DEY
        JR NZ,BACK                   BNE LOOP
        ;rest of program             BNE LOAD
                                     ;rest of program
```

Figure 13.12   Using double registers, or one timing loop inside another, for much longer time delays

Timing loops are used in many applications, among which are:

1. Timing control operations. The timing loop is started by an input signal, and an output signal is generated at the end of the timing period (traffic-lights timing).

2. Serial input or output. Bits from a byte are transferred from an input or output at a fixed rate which is much lower than the microprocessor clock rate. A timing loop must be used between each input or output step and the next to form the correct bit transfer rate.
3. Keyboard input. Mechanical key contacts usually bounce for a millisecond or so before finally closing. When a key is closed, a time delay is used to check the key value again before reading the value into the accumulator.
4. Forming pulses. A waveshape can be synthesised by outputting a byte followed by a time delay, followed by another byte. This can be used in the synthesis of music or speech.

Data-shifting instructions

A very large part of all microprocessor programming consists of instructions which shift data bytes from one memory or register to another memory address or register. The obvious examples are when a byte is copied from memory or from a port into the accumulator for processing, and where a byte, after some processing, is sent out from the accumulator to memory or to another port. In addition, however, it is very often necessary to preserve a byte value to use again after some intermediate operation, so that transfers between registers or between registers and memory are used much more frequently than simple input and output transfers.

A multiplication, for example, involves adding a byte to the same byte, shifted, so that storage is needed for the unshifted byte. Similarly, in a program of the 'traffic-lights' type, the byte which represents the state of the lights may have to be held temporarily in memory or in an unused register while the accumulator is used for the time delay steps.

A microprocessor which has a large number of registers, such as the Z80, can make use of otherwise idle registers to carry out this temporary storage. Transfers between registers are fast and need only a single-byte instruction, so that programmers tend to favour this type of storage, making use of the alternate register set of the Z80 if required.

Memory is also used for temporary storage, however, and this is a much more conventional way of storing bytes temporarily, because the amount of storage which is available in memory is always much greater than is available using the registers within the microprocessor. For this reason, most microprocessor chips contain instructions in their instruction set which allow memory to be accessed rapidly for this particular purpose. The section of memory which is used is given the name of *the stack*.

The stack

The stack is a section of memory, which can be chosen (except in the 6502) by the programmer to be reserved for the temporary storage of bytes. Many microprocessor programs that are intended for machine control make little or no use of stack memory. The use of the stack is generally more important in longer programs, particularly where fairly long computations are carried out between input and output steps. The principles which lie behind the use of a stack are, however, important and should be understood by anyone who will be involved in programming, whatever type of programming is intended.

The stack is simply a part of the normal RAM of the system, usually at the higher end of the address range of the RAM. The starting address of this stack memory must be available to the microprocessor in some way, and this is normally done by keeping the top-of-the-stack address in ROM, and loading it into the microprocessor when starting up.

The phrase, 'top-of-the-stack' is significant. When a memory address, for example 7FFFH (in an 8-bit processor), has been selected for use as the start of a stack, and built into the ROM of the system, each use of the stack will change the address of the next vacant memory. The stack address will be loaded from ROM into a register which is not normally directly accessible to the programmer, and which is called the 'stack pointer'.

This 16-bit register is used to maintain the address of the next free-byte in the stack, so that it is here that the address 7FFFH in our example will be stored from ROM at the time of switching on. When a byte is stored into the stack memory, it will be placed in the address 7FFFH which is held in the stack pointer, and the stack pointer register will then be decremented to 7FFEH, ready for another byte.

If, however, after a few more operations, the byte which was stored at 7FFFH is needed again, the instruction which fetches it will cause the stack pointer register to increment, so that it now addresses 7FFFH, and the byte is read out of the stack. Remember that the phrase 'read out of' must not be taken literally — the byte is still present in stack memory, but will be replaced by the next byte that is written in.

Stack order

Many descriptions of stack memory use the phrase 'top-of-the-stack' as if a byte were always stored in the top-of-stack memory address, with all the other bytes pushed down by one address number, as if in a shift register. This is very misleading, because such an operation would be slow, and is quite unnecessary. The stack pointer is the only register which is changed, and each use of the stack will either increment or decrement the stack pointer, so that the correct address in the stack memory is used.

263

If, for example, three bytes which we can label X, Y, and Z are stored in that order on the stack, the contents of the stack, and the contents of the stack pointer will look as in Figure 13.13.

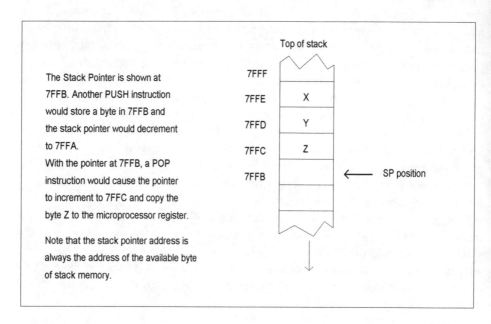

The Stack Pointer is shown at 7FFB. Another PUSH instruction would store a byte in 7FFB and the stack pointer would decrement to 7FFA.

With the pointer at 7FFB, a POP instruction would cause the pointer to increment to 7FFC and copy the byte Z to the microprocessor register.

Note that the stack pointer address is always the address of the available byte of stack memory.

Top of stack

7FFF
7FFE   X
7FFD   Y
7FFC   Z
7FFB   ← SP position

Figure 13.13   Using the stack

When the bytes are read back into the microprocessor, then because the stack pointer can only be incremented or decremented one step at a time, they must be read in opposite order, with byte Z, in this example, read first, since this is the byte in the address to which the stack-pointer points when it is incremented.

The 'last-in-first-out' order is strictly maintained, unlike the usual random-access use of RAM. This might seem to be a disadvantage, but it does allow a single-byte instruction to be used, as is usual when a byte is transferred from one register to another. The use of the stack is also very much faster than other memory transfers, because no addressing is needed — the stack pointer has the address ready for use. The phrase 'top-of-the-stack' therefore refers to the address of the last byte which was stored in the stack, and the word 'top' must not be taken literally.

Stack operations

The stack operations in assembler language, leaving aside the methods used for loading the stack pointer register, are PUSH (6502 and Z80) and either POP (Z80) or PULL (6502). PUSH means that the byte is pushed on to the stack at the

264

current address held in the stack pointer (top-of-the-stack); following a PUSH, the stack pointer is decremented to the next available address. POP or PULL means that a byte will be read, after incrementing the stack pointer, from the last stack address which was used for a PUSH. There may be a choice of what destination can be used.

The most frequently met use of the stack is for saving the contents of the registers. For example, a machine-control program which permits interrupts must also allow normal processing to continue after an interrupt has been dealt with. The usual method is to PUSH the contents of each register on to the stack, along with the program-counter address, at the start of the interrupt and then POP them off again at the end of the interrupt, so that operations can continue using the same data and at the correct address.

An example of this type of use is shown in Figure 13.14(a). The Z80 has instructions which allow each register pair to be placed on the stack or popped from it, and it is also possible to pop to a different destination, as illustrated in Figure 13.14(b). The 6502 has two PUSH and two PULL instructions, referring to the accumulator and the status register respectively; if the contents of other registers are to be saved, they must be first transferred to the accumulator.

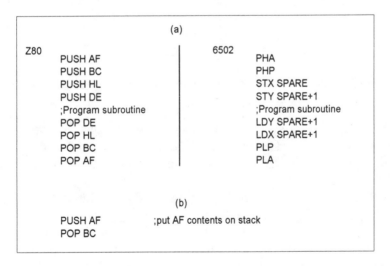

Figure 13.14   (a) Using stack commands to preserve register contents. This is done automatically when an interrupt occurs. (b) Registers must be popped in reverse order, otherwise contents will be exchanged.

265

NOTE: Memory locations can be used to save X and Y register contents for the 6502; this is faster then transferring data to the accumulator and placing the data on to the stack.

Subroutines and their uses

A subroutine is a piece of program which may be used more than once in the course of a longer main program. For example, consider a program which generates signals for switching traffic lights. The time between light changes is set by the time delay which is used between switching operations (Figure 13.15), and, as shown, this requires four time delay sections.

```
LIGHT A ON
DELAY
LIGHT B ON
DELAY
LIGHT A OFF
LIGHT B OFF
LIGHT C ON
DELAY
LIGHT B ON
DELAY
BACK TO START
```

Figure 13.15   A program scheme which can be more efficiently carried out using a subroutine

If the timing is to be symmetrical, meaning that the time for each section is equal, however, the time delay programs must be identical, and there is little point in writing the same steps several times. The solution is to make the time delay into a subroutine.

The difference between the original program and a subroutine version is shown for a few steps in Figure 13.16. In the original program, the time delay was started immediately after an output step, simply because the starting address of the time delay followed the last address of the switching routine. When a subroutine is used, the instruction CALL, followed by the subroutine start address, is used. This causes the time-delay subroutine to do its (time-consuming) thing, and then control returns to the address which follows the address of the CALL instruction.

```
        Z80                        6502

START:  LD A,01H          START:   LDA #01H
        LD(PORT),A                 STA PORT
        CALL DELAY                 JSR DELAY
        LD A,03H                   LDA #03H
        LD(PORT),A                 STA PORT
        CALL DELAY                 JSR DELAY
        LD A,04H                   LDA #04H
        LD(PORT),A                 STA PORT
        CALL DELAY                 JSR DELAY
        JR START                   JMP START
```

Figure 13.16   **Using a subroutine for a delay. This avoids having to program the delay steps more than once, assuming that the delay is for the same length of time. The delay routine is not illustrated**

The important feature of subroutines is that they can be called into action from any part of a program, and will return control to the address following the CALL instruction at the end of the subroutine. Each subroutine is entered by using the CALL (subroutine address) instruction, and each subroutine must end with a RET instruction, meaning return to the calling program. In the example of Figure 13.16, the delay routine could be any of the time delay routines that have been illustrated, with the RET (Z80) or RTS (6502) at the end of the routine.

NOTE that the codes of a subroutine can be placed anywhere, but it must not be possible for them to be run accidentally, so they are normally placed following an ending instruction. An alternative is to place all subroutines near the start of a program, with a bridging jump so that the subroutines are not executed in the normal course of incrementing the program counter.

All microprocessor types permit the use of subroutine calls, using the commands CALL (JSR for the 6502, mnemonic for Jump to Subroutine) and RET (RTS for 6502) in the instruction set. In addition, subroutines can be used only if a stack memory is available and the microprocessor stack pointer correctly set at the start of the program. This is because the CALL command automatically causes the content of the PC to be placed on the stack so that the program can return to the address after the CALL. In addition, the programmer will also need to write the subroutine starting with some PUSH instructions so that register contents can be preserved on return.

At the RET command, the contents of the stack are popped to restore the registers and to ensure the correct address for resuming the main program. If the RET command is omitted, the program will probably resume at the wrong place, with registers incorrectly loaded, and also with unwanted bytes on the stack — a recipe for disaster.

The examples which are noted in Figures 13.17 to 13.19 are of relatively simple subroutines which are used to a considerable extent in programs. Their inclusion is intended to give the reader some taste of the assembly languages of the 6502 and the Z80; obviously, working within the space of a single book of this type, a full treatment is impossible. In any case, excellent specialist books exist for the reader who wishes to continue the study of one assembly language.

The first program in Figure 13.17 will exchange data bytes between two memory addresses. It is important to remember that writing to a memory location destroys the byte which was previously present at that address, so that both bytes must be placed into registers before they can be exchanged.

```
Z80                            6502

        PUSH HL                        STX SAFE
        PUSH AF                        STY SAFE+1
        LD A,(FIRST)                   LDX FIRST
        LD L,A                         LDY SECOND
        LD A,(SECOND)                  STX SECOND
        LD(FIRST),A                    STY FIRST
        LD A,L                         LDY SAFE+1
        LD(SECOND),A                   LDX SAFE
        POP AF                         RTS
        POP HL
        RET
```

Figure 13.17   Exchanging data bytes in two memory positions

If we assume that the bytes are in consecutive memory addresses, then the use of indirect addressing is very convenient for the Z80; the 6502 can use indexed or absolute addressing. Only one method has been illustrated; there is seldom one single correct way of programming, though there is usually one way for a given problem of programming for the least use of memory, or the shortest time, or perhaps both. The set of routines in Figure 13.18 deal with timing, using loops of various lengths. Once again, the comparatively simple countdown steps which are used do not differ greatly from one microprocessor type to another, but the use of the double register of the Z80 makes for simpler programming.

The really large differences between the Z80 and the 6502 are well illustrated by the block shift programs in Figure 13.19, however. The aim is to move a block of bytes from one starting address to another, and the technique is to address the

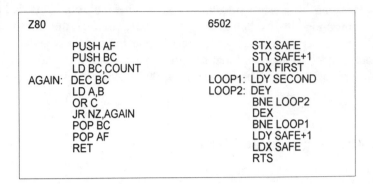

Figure 13.18   Timing routines for the two microprocessor types

```
Z80                                 6502

          PUSH AF                           STX SAFE
          PUSH BC                           STY SAFE+1
          LD BC,COUNT                       LDX FIRST
AGAIN:    DEC BC                   LOOP1:    LDY SECOND
          LD A,B                   LOOP2:    DEY
          OR C                               BNE LOOP2
          JR NZ,AGAIN                        DEX
          POP BC                             BNE LOOP1
          POP AF                             LDY SAFE+1
          RET                                LDX SAFE
                                             RTS
```

```
Z80                                 6502

          PUSH HL                           STX SAFE
          PUSH DE                           STY SAFE+1
          PUSH BC                           PHA
          LD HL, FIRST                      LDX #00
          LD DE, SECOND                     LDA COUNT
          LD BC,COUNT                       STA STORE
          LDIR                     LOOP:     LDA(X) FIRST
          LD A,B                             STA(X) SECOND
          POP BC                             INX
          POP DE                             DEC STORE
          POP HL                             BNE LOOP
          RET                                PLA
                                             LDY SAFE+1
                                             LDX SAFE
                                             RTS
```

Figure 13.19   Shifting a block of bytes from one starting address to another

first byte, read it, shift it, decrement a counter, increment both source and destination addresses, and repeat until the counter is zero.

The 6502 methods are more conventional in the sense that they show the individual steps that are needed, and which will have to be programmed for RISC microprocessor types.

The Z80 can make use of its block transfer instruction, which shifts from an address contained in the HL register pair to an address held in the DE register pair, using the BC pair as a counter. The decrementing of the counter and the incrementing of the address registers is carried out automatically. Corresponding instructions exist for decrementing both address registers.

Subroutines can be nested, meaning that another subroutine is called from within a subroutine that is being executed. Provided that the subroutines are correctly written this does not present a problem as long as the registers are saved on the stack in each routine, and that the stack space is sufficient to accommodate the amount of data that will be stored when the innermost level of subroutine is running.

Trace tables, registers and buses

A useful aid to programming is the use of trace tables. These show the contents of registers and the data on buses at each step in a program, and are particularly useful when the programming is complex, for example, when the stack is being used to transfer register contents. The full use of trace tables is too long a procedure for programs of more than a few instructions, but the exercise is useful, and is illustrated in Table 13.5.

Table 13.5   Using a set of trace tables on a simple routine. The characters xx mean that the value is unimportant

	IR	PC	SA	A	ALU	F	Comments
1	xx	7F00	xxxx	xx	xx	xx	Starting address in PC
2	BA	7F00	xxxx	xx	xx	xx	BA is first instruction byte
3	BA	7F01	xx14	xx	xx	xx	Fetch low byte of address STOR1
4	BA	7F02	1214	xx	xx	xx	Fetch high byte of address STOR1
5	BA	7F02	1214	06	xx	xx	Byte from address
6	DA	7F03	1214	06	xx	xx	Next instruction byte
7	DA	7F04	1215	06	xx	xx	Low byte of address STOR2
8	DA	7F05	2215	06	xx	xx	High byte of address STOR2
9	DA	7F05	2215	12	06	xx	Byte fetched, first to ALU
10	DA	7F05	2215	18	18	xx	Addition done
11	FA	7F06	2215	18	18	xx	Next instruction byte
12	FA	7F07	2216	18	18	xx	Low byte STOR3
13	FA	7F08	2316	18	18	xx	High byte STOR3
14	FA	7F08	2316	18	18	xx	Answer stored
15	00	7F09	xxxx	18	18	xx	End command

This shows the trace steps for the action of reading a byte from an address STOR1, adding to it a byte from address STOR2, and saving the result in an address STOR3. Note that displays of this type can be obtained using monitor programs.

In the table the abbreviations are:

IR   Instruction register (not usually available to a programmer)
PC   Program counter
SA   Store address (not usually available to a programmer)
A   Accumulator
ALU   Arithmetic and logic unit
F   Flag register

Exercise 13.1

Plan and develop a program for lighting and extinguishing a set of five LEDs. Each LED should light in turn, be on for about 1s, and be turned off when the next LED is turned on.

Exercise 13.2

Plan and develop a program that will switch over a reed-switch when the digits 35FC are typed on a hex keypad. Could this be used as part of a burglar alarm circuit?

Exercise 13.3

Plan and develop a program to operate a stepper motor, turning the shaft through 45° each 250ms.

Exercise 13.4

Plan and develop a program to control the speed of a DC motor, so that the numbers 1 to 10 on a keyboard correspond to speeds of, say, 100rpm to 1000rpm.

Exercise 13.5

Plan, write and test a program that will test a portion of RAM. This is to be done by writing alternate 1 and 0 bits for each byte, then reading this back to check that the same pattern exists. The action is then repeated with the bits reversed, so that 01010101 is used instead of 10101010, reading this back to check that the byte has been stored.

Exercise 13.6

Write a program to test a PIO, using the input and output lines connected together. Write a program to test a UART.

Exercise 13.7

Write a program to create a square wave from a PIO, using a frequency of 1kHz.

Test Questions

1. Give two reasons for using subroutines in a program. What stack actions must be included at the start of a subroutine?
2. The number CAH is stored at address 0400, and the numbers A5H and 1FH are stored in the following two locations. Draw a flowchart and write pseudo-language for a program that will add the first two numbers and add 1 to the third number if there is a carry from the sum of the first two. Assume that you are using an 8-bit processor.
3. The address 0400H contains an op-code 21H which will load the HL register of a Z80, and address 0403 contains the op-code 22H which will store the HL contents in memory. The address 0800H contains the byte AFH. Draw up a table of address bus and data bus contents and show the steps in loading the HL registers with the byte in 0800H and storing this byte in the address 0900H.
4. Write a delay routine for the 6502, using a stored number of FFH. Calculate the delay time for a 1MHz clock if the following number of clock cycles are needed for the instructions as follows:
    LDA 2   STA 4   DEC 6   BNE 2   RTS 5
5. Explain why it is relatively easy to re-write a Z80 program in assembler language so that the program will assemble and run on the 8086.
6. Explain the difference between a memory-oriented microprocessor and a register-oriented microprocessor. Into what class does the Intel 80486 fall?
7. Using the hex codes listed below for the 8086 processor, write machine code sets (operator and operands) that will perform the following on a 16-bit word stored in the AX (16-bit accumulator) register:
    (a) Set bit 3 of the register, leaving other bits unaffected
    (b) Perform a 2s complement on the register contents
    (c) Divide the register contents by 2 (ignore any borrow bit)

Action	Register	Code
Shift right	AX	D1 E8
Logical NOT	AX	F7 D0
Add immediate word	AX	05
OR immediate word	AX	0D

8. For the following program, draw up a trace table to show the contents of the PC, HL register pair, the accumulator and the memory addresses, all in hex, after each instruction has been completed. The MPU is the Z80 and the program shown here is part of a larger program. Address ADDR1 contains the byte 3FH and address ADDR3 contains the byte 6BH.

Address	Assembler code		Codes
7E7D	LD	HL,ADDR1	21 94 7E
7E80	LD	(ADDR2),HL	22 83 41
7E83	LD	HL,ADDR3	21 01 7F
7E86	LD	(ADDR4),HL	22 80 41

9. Construct a flowchart for adding two 16-bit numbers, using an 8-bit microprocessor. The result is stored in the address ANSWR. If there is a carry from the most significant bit, it should be used to set a bit in a memory location called CARRY.

10. Name one action relating to the use of the stack that is performed *automatically* when a subroutine is executed.

14   Microprocessor applications

Syllabus references: D03–3.2, M07

Microcontrollers

A microcontroller is a form of dedicated microcomputer, usually in the form of a single chip. *Dedicated* in this sense means that the microcomputer is intended to control one particular machine or system, and so its program is never altered except when the system is updated. The program can be in ROM or EPROM rather than in RAM as is used for desktop computers.

Another important requirement of a microcontroller is the use of ports. All but the simplest of controllers will need to control more than one action, and the control may need to be digital or, less usually now, analogue. For example, the amount of metal turned by a lathe may be controlled by a stepping motor, and this will need an output of one pulse (or set of pulses) for each incremental step. The speed of turning may be controlled by using an electric motor with a thyristor controller. Movement of the capstan so as to bring different tools into use can make use of a digit-code for each tool position. All of these signals will require the use of ports, and there will also be input signals such as measurement which must also use ports.

The essentials of a microcontroller are therefore a CPU along with ROM and RAM and ports on a single chip. There are many varieties of MPUs which answer to this description, because the market for dedicated microcontrollers is even larger than that for desktop computers. Tables 14.1 and 14.2 are of type numbers of a few well-known microcontroller chips using 4- and 8-bit data units. Most of the devices noted below exist in several different versions using different technology (NMOS, PMOS or CMOS) and with differing amount of ROM, RAM and ports. Several manufacturers supply identical designs which can be identified by the use of the same number range.

Table 14.1   4-bit controllers

Type designation	Manufacturer	Notes
HMCS45	Hitachi	NMOS, 2K x 10 ROM, 160 x 4 RAM, 44 I/O
MN1405	Matsushita	NMOS, 2048 x 8 ROM, 128 x 4 RAM
MN1564	Matsushita	NMOS, 4096 x 8 ROM, 256 x 4 RAM, 52 I/O
MC141200	Motorola	PMOS, 1024 x 8 ROM, 64 x 4 RAM, 25 I/O
COP422	National	NMOS, 1024 x 8 ROM, 64 x 4 RAM, 16 I/O
µPD75316	NEC	16K ROM, LCD driver
MSM5845	OKI	CMOS, 1280 x 8 ROM, 64 x 4 RAM, 30 I/O
MM78	Rockwell	PMOS, 2048 x 8 ROM, 128 x 4 RAM, 31 I/O
ETC9445	SGS-Thompson	CMOS, 2K ROM, 128 x 4 RAM
TMS1100	Texas	PMOS, 2048 x 8 ROM, 128 x 4 RAM, 25 I/O

Table 14.2   8-bit controllers

Type designation	Manufacturer	Notes
HD6805X1	Hitachi	4K ROM, 56 I/O, timer, serial port
8022	Intel	2K ROM, 64 x 8 RAM, ADC
8751	Intel	4K EPROM, 128 x 8 RAM
M5L8049	Mitsubishi	2K ROM, 128 x 8 RAM
MC6801	Motorola	2K ROM, 128 x 8 RAM
MC6805P6	Motorola	1.8K ROM, 20 I/O, timer
µPD78224	NEC	16K ROM, 640 x 8 RAM, 63 I/O
MAB8051	Philips	4K ROM, 128 x 8 RAM
R6500/1	Rockwell	2K ROM, 64 x 8 RAM
M3876	SGS-Thompson	6K ROM, 32 I/O
ST9040	SGS-Thompson	16K ROM/EPROM, 7 I/O, ADC, serial
Z8671	SGS-Thompson	BASIC controller with debug software in ROM
SAB 8048	Siemens	1K ROM, 64 x 8 RAM
TMP8748	Toshiba	1K EPROM, 64 x 8 RAM
Z8600	Zilog	2K ROM, 144 x 8 RAM

Real-time operation

The use of a microprocessor in real-time is the most demanding of applications. Real-time means that the processing must be completed in the time that is allocated by other processes. For example, if a microprocessor is used to control the speed of a petrol engine there must be no appreciable delay between feeding in the data and obtaining the controlling signals — it would not be acceptable to have even a 0.1 second delay in an application like this. Similarly if a microprocessor system is used in EPOS (electronic point of sale) equipment, it would not be acceptable to notify stores that an item was out of stock several hours after the last of that item had been sold.

The main problem of using a microprocessor for real-time applications is speed of response. The conventional microprocessor is serial, with the instructions being executed one after the other rather than in parallel. This leads to programs being very long, with many thousands of steps, and this in turn means that the time can be significant. This problem is not confined to the microprocessor — conversion of analogue signals into digital requires very fast converters, as Chapter 5 has indicated.

Real-time control is easiest when the time available is not too short, and when inputs and outputs can both be in digital form. When analogue-digital conversion is needed, the time that this requires will require faster performance from the microprocessor to carry out the other steps in the control action.

Consider, as an example, the problem of controlling motor speed. Imagine that the motor shaft carries a rotary encoder, so that a set of pulses whose frequency is proportional to motor speed is available. The control of motor speed can be made by using a thyristor circuit which also can be pulsed. The micro-controller has to set the motor speed to a level determined by a 64-level input signal, and to maintain the set speed as far as possible independently of loading.

Before we can think of solution, we need to look in more detail at these inputs and outputs. The input from the rotary encoder can be sampled to find the time between pulses. It seems reasonable to sample only once per revolution or perhaps once per five or ten revolutions; we might even make the encoder a very simple one that cuts a light beam only once per revolution, but this is unlikely to be suitable if the motor has to be controlled down to a very low speed.

The output pulses to the thyristor circuit also need some investigating. A design cannot be attempted until we know how pulse rate is related to motor speed, and if this is not known it will have to be found by practical experiment. Finally, we need to know what will be used for setting motor speed — a potentiometer (analogue output) or a multi-way switch (digital output).

Once the facts are known, we can start on design work, and an outline is illustrated in Figure 14.1. The input pulses will provide a time reading, and the

Find time T between input pulses

Find time t set by speed setting

Compare T and t, find difference

Use difference to change output pulse rate

Figure 14.1   A first attempt at a program plan for a proposed motor-speed control

speed setting, imagined as a switch, can be used along with a table in ROM, to provide another time setting. The difference between these two will be used as a correction number, and will control the generation of pulses for the output.

This leaves some questions unanswered. What pulse rate will be used for the thyristor output when the unit is first switched on? How do we alter the rate of output pulses in response to the input information? The answers depend on the method that is used to create the output pulses, and the obvious answer for the first portion is that the switch-speed control setting looks as good a starting point as any when the unit is switched on.

A pulse is created by sending a 1 bit to a port, waiting for a fixed time, then sending a 0 bit to the same port and following this also with a time delay. In pseudo-language this is:

```
START: Send out 1
       delay
       Send out 0
       delay
       Jump to START
```

The frequency of this set of pulses depends on the number that is used in the counter that will determine the delay setting, and if this number is stored in memory or in a register it can clearly be changed to alter the frequency of the pulses.

We can imagine then that at start-up, the program performs the subroutine that reads the setting input and finds the appropriate setting number from the ROM. This number will then be used in the delay memory address as a way of setting the pulse output frequency. When the shaft starts to spin and a time figure is available, this is compared with the setting of time, and the difference (a signed number) can be added to from the time number that is held in memory and used as a delay. When the shaft is rotating slowly, the time number from the encoder will be large, and when this is subtracted from the setting number the result will

277

be negative, so that the time setting for the delay will be reduced. If the shaft rotates too fast, the time difference number will be positive and the delay time will be increased, lowering the output frequency and reducing the motor speed.

This will be the main loop, calling the delay subroutine twice for each complete pulse, and the effect of running this program is to make the controller part of a negative-feedback circuit. A useful refinement would be to ensure that the initial setting is not altered until the motor speed become established. This might be controlled by the time between pulses, or by the time that has elapsed since the motor was switched on.

These steps illustrate the type of comparisons and number storage actions that are commonly used in machine control.

A typical microcontroller — the TMS1000

The TMS1000 (Texas Instruments) is part of a family of 4-bit controllers which are intended for use with mass-produced machines that can range from toys through musical door chimes to simple industrial process control. Their low cost

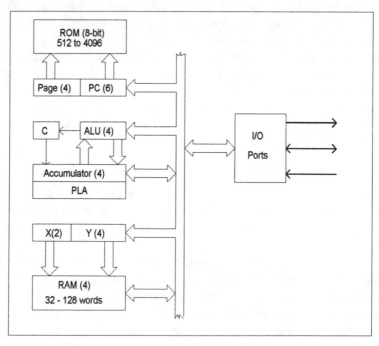

Figure 14.2 **The block diagram for the TMS1000 controller. The ROM and RAM sizes depend on the version used**

has made them the most widely used of all 4-bit controllers in spite of intense competition from more modern designs.

Figure 14.2 shows a block diagram of the internal organisation of this chip, which follows the architecture of the simpler varieties of 8-bit computer chips. The TMS1000 family of chips are designed very much like 4-bit calculator chips, using separate program ROM and data RAM spaces. The four-bit ALU handles arithmetic and logic, with a carry flip-flop. The RAM can be addressed by both X (2 or 3-bit) and Y (4-bit) registers, using this split addressing scheme so that a 4-bit bus can be used.

The program counter uses 6-bits to address 64 bytes (8-bit) of ROM program memory along with a 4-bit page register which allows up to 16 'pages' of 64 bytes to be used; a total of 1,024 bytes. The chips that can use larger amounts of ROM do so by using two extra flag bits, termed 'chapter' bits, with each chapter number selecting a page of 1,024 bytes.

The TMS1000 family chips do not provide for a stack, but a buffer register, the SAVE register, can be used to save the PC contents, page number and flags when a subroutine is called.

There is a clock generator on the chip, and the block marked I/O can provide various numbers of I/O lines according to the model number. I/O line numbers range from 4 to 33. A few types add a driver for a fluorescent display as used in calculators and the programmable logic array (PLA) on the chip can be used to provide segment decoding for displays. The PMOS versions use signal levels that are not TTL compatible, but the NMOS and CMOS devices use TTL compatible signals.

Figure 14.3 shows the pin-out for the TMS1000 family that use a 28-pin DIL pin-out; the TMS1200 family of similar devices use a 40-pin DIL package, and the specialised TMS1098 and TMS1099 chips (no ROM, used for prototype design) use a 64-lead DIL package.

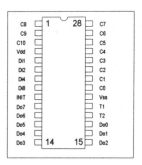

Figure 14.3   **The pin-out for the 28-pin DIL versions of TMS1000**

279

The 11 control output pins can be set or reset individually, and can be addressed by using the Y-register. The eight Dout pins are driven through the PLA portion using data derived from the accumulator. The programming of the PLA chip can be performed to customer requirements since this is a masked ROM. The four Din pins can be used with keyboards or with individual switches.

The instruction set consists of 43 actions (54 for a few types) consisting of arithmetic and logic, data transfer, branch and jump and a CALL instruction for a subroutine — because there is no stack there can be no nesting of subroutines though a few varieties provide extra save registers to allow two- or three-level nesting.

The clock frequency is set by connecting timing components between the T1 and T2 pins, and the usual operating frequency is in the 300–500kHz range, though higher clock rates can be used with the CMOS types.

Using a computer for control

Process control can be achieved by using a desktop computer along with suitable software. The advantages of this scheme are that the control can be much more flexible, the resources of memory can be much larger, and that the programming can more easily be done using a higher-level language (such as BASIC or C) rather than in laborious assembler language. In addition, the screen display and the printer output, along with keyboard and mouse inputs, can be incorporated into the control actions.

The main disadvantage is that the usual type of desktop machine is not richly endowed with inputs and outputs; the usual provision is one parallel and two serial ports, with no A/D or D/A converters. Another point is that the programming of a desktop machine can be too easily changed, making the system vulnerable to interference, accidental or deliberate. In addition, the desktop machine may be too slow for real-time work because its operating system and programming has been written in a high-level language rather than in fast and efficient assembler language.

The PC type of machine can be fitted with internal interfacing cards which provide for as many input and output ports as could be achieved using a dedicated microcontroller. The R-S Components catalogue lists boards manufactured by Arcom Ltd. which provide up to 40 parallel I/O lines for process control purposes.

Monitor programs

The word 'monitor' can be used in two very different senses. The hardware meaning is that of a VDU — a display system. Programmers, by contrast, take the word to mean a program that is used in connection with developing other programs. Typical actions of a monitor program include memory and register dumps, single stepping, subroutine skip, and breakpoint setting.

A *memory dump* is a set of readings of memory addresses and contents which can be made at any starting address and continuing for as many bytes of data as are of interest. Conventionally, monitors display a memory dump in hex, but some provide for addresses in hex and data in ASCII codes. You would use a memory dump to check that a program was using memory in the way that you expected.

Register dumps are even more useful. A *register dump* will display on the screen the contents of all the programmable registers of the MPU at any stage in a program, allowing you to check that the register contents are what you would expect from a trace table (see Chapter 13). You can also use a monitor dump to find what signals exist at I/O ports.

These dumps are used in conjunction with *single-stepping*, so that the program can be run one step at a time from any starting address, and the memory and register contents displayed for each step. You would normally carry out a single-step analysis on a part of the program that you suspected of causing problems, because to single-step a complete program would be very time-consuming. This type of action is called 'debugging' and a useful way of remembering it is that a fault in a program is called a bug, finding the fault is called debugging, and the cause of the fault is called a programmer.

Subroutine skip allows you to single-step until a subroutine is called, and then move through the subroutine at normal speed. This is useful if you have used tested subroutines in a program and you know that they are unlikely to be at fault.

The other common option is setting breakpoints. A *breakpoint* is an address where execution of a program at normal speed can be made to stop. This allows you to run a program normally up to a point where you have found that problems start. At the breakpoint you can then use a memory and register dump, or you can single-step to find precisely where the trouble starts.

Monitors usually provide for many other actions, even for writing and changing pieces of code, but the actions listed above are the most used and most valuable. Monitors can be obtained for machine-code or for higher-level languages, but the monitor must match the processor (for machine code) and the language and computer for the higher-level languages.

The use of a monitor with a desktop computer makes it possible to carry out the programming for a dedicated controller using the computer. A form of program called a 'cross-assembler' allows you to write programs in the assembly language of the controller, and a matching monitor allows you to check the action of the code. This can speed up the development of a controller circuit very considerably because it is not necessary to create an EPROM until you are fairly sure that the machine code is suitable. In many cases, the monitor can be relied on sufficiently to allow you to save the code on disk and have masked ROM made from this code directly.

Exercise 14.1.

Using a setup as indicated in the sketch, Figure 14.4, use the computer to monitor and control the temperature in the closed box.

Exercise 14.2.

Using a trolley driven by a stepping motor, use a microcomputer to control the distance for which the trolley will move.

Exercise 14.3.

Using a DC motor with thyristor or field-current control, devise a program to control the motor speed.

Test Questions

1. Compare real-time and artificial-time applications in terms of the need for (a) speed, (b) memory, (c) disk use.
2. What are the main differences between a general-purpose microcomputer and a dedicated microcomputer?
3. Outline how you could use a microcomputer to control the water-on, water heat, wash, spin, and drain cycles of a dishwasher for commercial use?
4. Explain how you could use a microcomputer to synthesise a 100Hz sawtooth wave.
5. Outline how you would use a microcomputer to program a microcontroller chip.

15   The PC machine

Syllabus references: B1–A, B1–C, B2–C, I1, I2, I3, I4

The DOS

Any computer that is intended for serious use must be able to store data on disks, including a hard drive as well as on floppy disks. The use of magnetic disks requires software that will find space for data and maintain a file that keeps a record of how each file is positioned on the disk. This form of software is called a disk operating system, or DOS. The DOS is usually extended to a large number of other actions, such as file copying, printer control, and monitor control. Figure 15.1 illustrates the relationship of DOS to the other parts of a working system from the user at the keyboard to the hardware inside the computer.

An Apple machine, named *Lisa*, pioneered for small computers the type of operating system that is now called GUI (graphical user interface) but which was formerly known as WIMP (windows, icon, mouse programming). In this form of system, developed originally by the Xerox Corporation, the screen showed programs and data files as pictures, called icons, and actions on programs and other files were carried out by using visual methods in conjunction with a mouse. The mouse is a small trolley whose driver software ensures that movement of the mouse causes a cursor to move on the monitor screen. The mouse is also

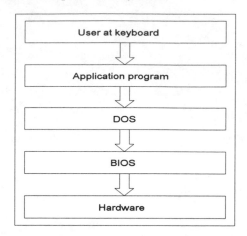

Figure 15.1   How the DOS fits into the layers of control

equipped with a button (one of two or three) which carries out the action of the RETURN or ENTER key.

In a system of this type, a program can be run by placing the cursor over it, using the mouse, and clicking the mouse button. The user does not need to remember or to type the name of the program, so that mis-spelling or mis-typing cannot cause any problems. The system provides also for an action called *dragging*, in which the cursor is placed on an icon, and the mouse moved with the button held down. Using dragging, a file can be copied to another directory by dragging its icon to that place, and a file can be deleted by dragging it to an icon of a waste-paper basket. The GUI fits into the diagram of Figure 15.1 as a layer between the application program and the DOS.

The Lisa was not a commercial success, but its successor, the *Macintosh* (named after a US variety of apple) was very successful and influential. The PC type of machine can use the Microsoft Windows system which provides these facilities, though DOS is still used as the underlying operating system with Windows providing a user interface or *front end*.

Base memory, expanded memory and extended memory

The PC machine, named after the IBM PC which first appeared in 1980, has evolved to become the leading machine for business uses, rivalled only by the Apple Macintosh. The reason for this supremacy is that it has allowed continuity of software from the outset, so that the oldest PC programs will still run on the modern PC, though the most recent programs will not necessarily run on an old PC. Though there are several hundreds of manufacturers making PC machines, all follow the same design pattern and will accept the same software. The competition among manufacturers has resulted in prices being steadily reduced over the life of this design. The features that mark out a PC machine are:

1.   The use of an Intel or Intel-compatible chip, such as the 80486.
2.   The ability to use MS-DOS as the operating system.
3.   The use of a keyboard that is connected through an asynchronous serial link.
4.   The use of memory arranged with ROM at the upper part of the first megabyte.

Figure 15.2 shows the standard arrangement of memory in a PC machine — modern PC machines continue the memory addresses beyond FFFFFH, and video boards can contain 1Mbyte or more of additional RAM so as to be able to use high-resolution graphics.

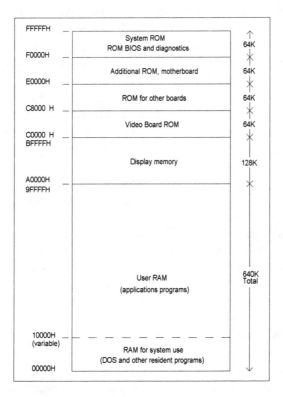

Figure 15.2   **The standard PC memory map, showing the base memory of 640K**

Because the older 8-bit microprocessors could work with a maximum of a 64Kbyte memory space, the original IBM PC design was organised so as to use memory in units of 64Kbyte. For the 1Mbyte of memory that this machine could address, 640Kbyte of memory was available as RAM for program use, with the remainder of the addresses reserved for system and other ROM chips. This 640Kbyte of RAM, using the lower portion of the available memory addresses, is known as *base memory*. The arrangement of the memory into ten segments of 64Kbyte each made it easy to adapt programs that had been written for older 8080-based machines to the new PC. Unfortunately, it makes it very difficult to use large program files, as compared to a machine with 'flat' addressing (all addresses from 00000H to the limit of memory available for RAM in a continuous set). In the early 1980s several machines were manufactured with less than 640K of base memory, and *expansion base memory* could be added to make the total RAM up to 640K.

The original PC machines used PC-DOS or MS-DOS (PC-DOS for the IBM machine, MS-DOS for others), which was tailored to the use of a 640Kbyte memory. When chips such as the 80286 and 80386 became available, which could address considerably more than 1Mbyte of RAM, the 640Kbyte limit for programs remained because this limit was now set by the operating system rather than by the addressing limits of the chip. No attempt was made to extend the PC-DOS or MS-DOS system, because by that time new ideas on operating systems were beginning to take hold. Machines which used different operating systems could use as much memory as their microprocessor chips could control, but lacked the advantage of being able to use the huge range of programs that had been written for PC-DOS and MS-DOS machines.

The success of the Macintosh (using the Motorola 68000 family of microprocessors) indicated that the future of microcomputers lay in the use of a GUI operating system rather than with developing MS-DOS or PC-DOS. Windows allows the use of as much RAM as can be physically provided, typically 64Mbyte or more, along with *multi-programming* (the ability to run more than one program at a time) and easy use, so that by 1992 the use of programs running directly under DOS was declining compared to the use of programs that were written to take full advantage of Windows. Older programs can still be run using Windows, so that the compatibility that has always been a feature of the PC machines is maintained. The importance of this is emphasised by the fact that the value of software for a PC is often three times as much as the total value of the hardware.

As programs, particularly CAD (computer aided design) packages and, later, spreadsheets, became larger, the 640K limit had to be broken. Before the use of Windows became established, there were two ways in which the 640Kbyte RAM barrier on a PC machine could be broken. One was to use additional memory in

64Kbyte segments, switching from one segment to another. This scheme was known as *expanded memory*. The other was to use the chips from 80386 onwards in a different mode, ignoring 64Kbyte segments. This scheme was called *extended memory*. The Windows system uses extended memory, and expanded memory is now obsolete. Modern machines provide for a minimum of 4Mbyte on the main circuit board (*motherboard*), with provision for using up to 128Mbyte. The use of expansion boards for adding memory is now unusual, as such memory cannot be accessed fast enough for modern machines (because of the lower clock speed used on the expansion bus). The version of Windows known as *Windows-95* dispenses with the need to start MS-DOS separately so that references that follow to the use of this system apply to machines which have not been upgraded to Windows 95 or which are using older DOS programs.

Note that the use by MS-DOS of *any* memory above the 640Kbyte limit requires special *driver* software. Windows 3.1 requires a driver called HIMEM.SYS to manage extended memory, and another driver called EMM386.SYS is used when the use of some odd portions of memory is required, or the use of expanded memory. On modern machines, EMM386 is often used to allow the use of some of the 64K blocks of memory in the lowest 1Mbyte that are otherwise unused.

PC video graphics — MDA, CGA, EGA, VGA, SVGA

The IBM PC was originally conceived as a home computer, but its text-only video adapter, the monochrome display adapter (MDA) could not possibly produce the graphics that were essential for games, and certainly not in colour. The design of the PC provided for additional circuit cards to be plugged into holders (called slots) in the main board (the motherboard) and IBM developed a colour graphics card, the colour graphics adapter or CGA card. The CGA card was an attempt to provide some colour graphics capabilities, but it was not well matched to the capabilities of the machine, and certainly did not provide as good a display as most games machines of the time. The card allowed two forms of screen display, a text screen for the display of text characters only, and a set of graphics screens with different resolution capabilities. The board was intended for a maximum resolution of 640 dots across by 200 down the screen, and only when two colours were being used, corresponding to a monochrome display. In its very limited four-colour display (and these four include black and white), the resolution was only 320 dots across by 200 down.

Even when colour was ignored, the CGA card was quite unsuited for either serious or games use, because its text characters were formed from an 8 x 8 set of dots rather than the superb 15 x 9 of MDA. This made the characters look

287

blurred, coarse and poorly-shaped, and quite unworthy of the machine, and even when a CGA card is used in a graphics mode, the resolution is still low. CGA cards require a monitor which uses a 15kHz horizontal scan rate and digital RGB signals. CGA boards are never used on machines intended for serious business uses.

The CGA card brought one important new principle, the use of 'screens', meaning different resolution modes that could be switched by software. All subsequent cards have featured a text mode and several graphics modes that can use various resolutions and choices of colours. In addition, the codes for several screen displays can be held in RAM on the video card so as to allow rapid switching from one screen 'page' to another.

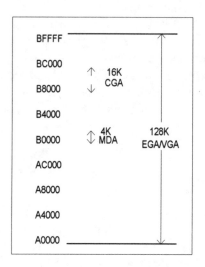

Figure 15.3   The use of video memory by MDA, CGA and EGA/VGA boards

IBM then developed a new graphics card called EGA, enhanced graphics adapter, which relied on increasing the horizontal scanning rate from the 15kHz of CGA to around 22kHz so as to permit a resolution of 640 x 350 dots. In addition, an analogue video signal is used, so that the choice of colours is greater. The normal range of colours is 16, but the 16 colours that can be displayed can be selected from a set of 64, so that 16 is the practical limit to the number of colours that can be displayed on one screen, rather than the absolute limit to the number of colours. In addition, the EGA board could display monochrome text on a 15 x 9 matrix, monochrome graphics of 720 x 350 resolution, and also CGA-type colour displays of 320 x 200 resolution.

When IBM introduced their PS/2 machines in 1987, they broke to a considerable extent with the compatibility that had persisted from the early days of the PC machines (8088 and 8086 chips) and extended to the AT machines (80286 and 80386 chips). The PS/2 machines featured two new video display systems, MCGA (multi-colour graphics array) and VGA (video graphics array), and a new range of monitors which were analogue only.

The VGA card has set the standard that business software has followed. VGA permits full compatibility with MDA, CGA, EGA and MCGA. In addition it adds displays of 640 x 480 16-colour graphics and 720 x 400 colour graphics, using a 9 x 16 grid for characters with colour. This card requires a high standard of monitor if colour is required, and the use of a mono monitor can be restricted to a few modes unless an analogue monitor is used. The VGA card is universally used on PC machines at the time of writing.

Enhanced VGA or Super VGA (SVGA) cards are also available, and are rapidly setting new standards. These cards can provide 640 x 480 resolution in 256 colours or even in 16 million colours, 800 x 600 in 16 colours and even 1048 x 768 or more in 4 colours or 16 colours. There is a trade-off between resolution and number of colours, because increasing either requires more memory for video, and the amount built into the video card is limited, often to 256Kbyte. Unless the card contains additional memory sockets that allow you to upgrade its memory to as much as 1Mbyte the higher resolutions must inevitably be restricted to smaller numbers of colours at any given time (though the colours used on a screen can be picked from a large range). The number of colours corresponds to the number of digital bits used to code colour, and it does not imply that your monitor can display such a range of colour or your eye detect any differences between them. A memory of 512Kbyte is adequate for most purposes unless you want to use more than 256 colours.

Screen displays generally follow a method called *APA*, all points addressable, meaning that each point on the screen corresponds to an address in video memory. The screen points are called *pixels* (picture cells), and the pixel is the smallest size of unit that can be displayed. The number of bits that can be stored at each video address determines the choice of colours that is available. For example, using one bit per pixel allows for two colours, usually black and white. With two bits, a choice of four colours becomes available, and four bits permits the standard EGA/VGA set of 16 colours. 8-bit storage allows 256 colours, and allocations of up to 24-bits are used for SVGA systems.

The mouse

The mouse, Figure 15.4, provides another way in which information is fed into the computer, and for graphics and drawing (CAD) programs, along with many

desk-top publishing (DTP) programs, is the only really serious way of working with such programs. A mouse (or an equivalent such as a pen or tracker-ball) is *essential* for using Windows or any other GUI.

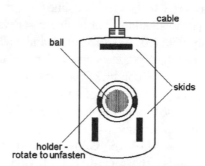

Figure 15.4   The mouse as seen from the underside

The mouse movement consists of a heavy metal ball, coated with a synthetic rubber skin, which can be rotated when the mouse is moved. The movement of the ball is transmitted to small rollers Figure 15.5 and then sensed in various ways, depending on the make of the mouse, such as magnetic changes, or by light reflection, and the signals that indicate the movement are returned to the computer to be used in moving the cursor.

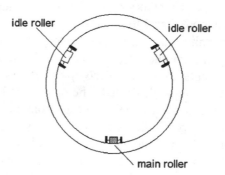

Figure 15.5   Mouse ball and rollers, used to sense movement

The mouse action depends on software being present, and this can cause a considerable amount of confusion. This nowadays is usually a program called MOUSE.COM which is run in the AUTOEXEC.BAT file (see below). This program remains in the memory of the machine until you switch off, so that you do not need to re-activate the mouse each time you change to using another program. If you use the Windows system, the mouse action is catered for automatically.

Some computers are provided with a *mouse port* on the motherboard, or a mouse port may be provided as an add-on card when you buy a bus mouse. It is more common to find that a mouse will be supplied for serial port connection. The advantage of using a serial mouse is that it is more likely to be possible to connect a serial mouse to any variety of PC machine than the bus mouse, because virtually any PC will have a serial port, not all have a mouse port. The snag is that you may very well need the serial port for other actions, such as the use of a modem or for connections between computers, so that it is an advantage to opt for more than one serial port when you are considering port provisions. In addition, the use of a serial mouse can interfere with the timekeeping of the internal real-time clock.

Note that the two mouse types are not interchangeable — the signals that pass between mouse and computer are quite different, and the connectors can be different as well. The bus mouse often uses a nine-pin D connector, and the serial mouse can use either a 9-pin D or a 25-pin D connector. The connections to the 9-pin D-plugs for a serial mouse are not the same as those for a bus mouse, and when a serial mouse is fitted with a 9-pin D connector it is usually packaged with a 9-to-25 pin adapter as well so that it can be used on serial port which terminates in a 25-pin socket.

Mouse maintenance

The ball should be cleaned at intervals, avoiding solvents — spectacle lens cleaner and a soft cloth are preferred. Clean also the rollers that can be seen inside the mouse casing when the ball is removed. These rollers are quite difficult to clean, and the best way is to wrap some cloth (a handkerchief will do) around the points of tweezers, moisten this, and rub it across each roller, moving it sideways a few times, then pulling or pushing to rotate the roller slightly and then rubbing sideways again. Dirty rollers are a much more common source of trouble than a dirty ball; typical symptoms are erratic or 'sticky' movement of the mouse cursor.

Booting up

When you boot up (switch on) the PC machine, following the self-test sequence called POST (power-on self test), files are read and actions performed in a never-changing sequence. The POST routines are held in the ROM BIOS. The first part of this uses the information held in the CMOS RAM to set up the drives and correct the clock. The second part makes use of information in a file called CONFIG.SYS to set up memory-resident pieces of code that determine some fundamental operating conditions. The third part of the sequence is reading in MS-DOS, and this is followed by reading and acting on another file called AUTOEXEC.BAT which also sets up memory resident codes. Before MS-DOS is read in, the machine checks for the presence of a floppy disk in the A drive, and if a disk is present, MS-DOS will be read from this disk. If the disk in the drive is not one that contains the MS-DOS system tracks, an error message will be delivered and the machine will wait for a suitable disk to be inserted. If there is no floppy disk in the A drive, the machine will read the MS-DOS system from the hard drive. This is why there must be no disk in the A drive when you boot up a machine that uses a hard drive.

Note that the separation of BIOS routines from DOS routines makes it easier to write applications programs that will work with any computer using the same DOS, because all the routines that are specific to the hardware can be placed in the BIOS.

Before any files are loaded, it is possible to run a SETUP routine from the ROM. This is done by holding down a specified key (such as F1) or key combination (such as Ctrl-Alt-Esc) when the machine is switched on. The SETUP screen will allow you to correct the clock and calendar, specify hard or floppy disk parameters, notify size of memory and type of video card and allow the use of shadowing ROM into RAM.

There is not much rhyme nor reason about which actions are performed in CONFIG.SYS and which in AUTOEXEC.BAT, and the trend in the last few years has been to use CONFIG.SYS only for essentials, with AUTOEXEC.BAT used for the more optional setup commands. The general rule is that files whose extension letters are SYS need to be dealt with in the CONFIG.SYS file.

The important point about CONFIG.SYS and AUTOEXEC.BAT is that both are *plain text files* which you can write and/or modify. When you change CONFIG.SYS the new version will take effect only after you re-boot the computer, but if you alter AUTOEXEC.BAT, you can make it carry out its actions by typing the name AUTOEXEC and pressing the ENTER key, just as you would do for any other command.

The control of the PC machine is done by way of signals that are sent from the operating system to a set of routines held in ROM. Because machines may be differently designed, these ROM routines must be tailored to the machine, and the routines are called collectively the BIOS, basic input output system. Two software files are used to pass the signals from the operating system to the BIOS, and these are known as IO.SYS and MSDOS.SYS (or corresponding names such as BIO.SYS and IBM.SYS). In addition, some operating system commands make direct use of routines in the BIOS ROM for faster action, and another time-saving technique is to copy (shadow) the contents of the BIOS chip into RAM because access to RAM can be much faster than to ROM. See Figure 15.1 for the layers of command diagram.

The sequence of DOS command control is therefore:
1. A command word is typed and the RETURN key pressed.
2. This locates a routine in MS-DOS, sending signals to addresses in IO.SYS and MSDOS.SYS.
3. The SYS files locate and activate routines in the BIOS.
4. The code in the BIOS is passed to the microprocessor.

Note that applications programs are almost always designed to run on a specific operating system. For example, if you buy the word-processor called Microsoft Word you must specify whether it is for the PC machine or the Macintosh, and some programs can be bought for other operating systems such as Unix or Pick, which are rarely used on microcomputers. The usual choice for PC programs at the time of writing is for DOS, Windows, or the IBM operating system called OS2. The most recent version of Windows, Windows 95, allows you to ignore MS-DOS and the CONFIG.SYS and AUTOEXEC.BAT files, because these files (which still exist) are controlled by Windows.

Programs for MS-DOS use filenames that can consist of three parts. The main filename can be of up to eight characters, and this can be preceded by a path of up to 64 characters that shows the progression from the root directory (see later) to the file position. An extension of up to three characters can follow the main filename, separated by a full-stop. The extension letters indicate the type of file, and the most important extensions are COM, EXE and BAT. COM is used for a short program that fits into one 64Kbyte segment. EXE is used for a longer program that will require several segments of memory, and BAT signifies a batch file that consists of DOS commands. Any of these files can be run by typing the main name alone, omitting the extension. The SYS extension is used for specialised purposes (see later). Other extensions are not standardised, so that TXT or DOC, for example, can be used for text files. Windows 95 allows filenames of up to 255 characters to be used, and when you use such names, an *alias* name conforming to the older rules is also stored so that files can be read by older software.

The CONFIG.SYS file

No two computers are completely alike, nor do any two computer owners use their machines in the same ways. Because of this, details of these differences are stored in two files that the computer reads when it is booted up. These files make sure that your computer is set up correctly for such things as the type of keyboard

(English, French, German ...) and that any programs that you want to keep in the memory are put there. One example of a program you would need to keep in the memory is the driver program for a mouse.

A very simple CONFIG.SYS file for a PC, such as might be created for you when MS-DOS was installed, might consist only of the lines:

    files=20
    buffers=20
    country=044

and it's likely that you would want to use at least this number of lines, probably many more, and probably with different entries for a different machine, particularly when memory has to be organised on a 386 or 486 type of machine.

In this example, the *files=20* line allows up to 20 files to be open simultaneously. This may seem large, but it is needed to allow for the way that some programs handle their files, working with several files at once and also creating temporary files. The *buffers=20* line allows 20 memory buffers to be maintained for files, speeding up filing actions. The *country=044* line sets constants for the UK, such as the use of the £ sign for currency, commas as number separators (as in 1,225,617) and dot as decimal point, and the day-month-year date format. The code 044 is the international telephone code for the UK.

MS-DOS 6.22 creates a more elaborate CONFIG.SYS file when it is installed, and until you have some experience with CONFIG.SYS files you should not change the file unless something *must* be changed, such as the keyboard type. The CONFIG.SYS file that is created when MS-DOS is installed will usually be satisfactory, because MS-DOS 6.22 can detect the type of monitor system, keyboard, language and other settings that you are using and it alters the CONFIG.SYS file to suit.

You can alter the CONFIG.SYS file of the computer using the MS-DOS editor (or any other text editor program). Do *not* use a word-processor unless you know how to use it to create the type of files called ASCII. Remember that changes that you make in this way to a CONFIG.SYS file have no effect until you re-boot the machine with the new version of the file. Always keep a copy of the older version when you change CONFIG.SYS, in case the new version causes problems.

Other command lines that you can find in the CONFIG.SYS file are *Break*, *Lastdrive* and *Device*. The Break line is seldom used nowadays, but if it is present it will be in the form *Break=On*, allowing the Ctrl-Break keys (pressed together) to interrupt disk operations. The default is that Break is off, so that disk actions cannot be interrupted. Lastdrive is used in a form such as *Lastdrive=K* to indicate the last letter that can be used as a drive letter. The default last drive letter is E, but if you need more drive letters (for network uses or because of devices that simulate drives) the Lastdrive line can allocate more letters.

The most important lines of the CONFIG.SYS file start with Device, and are used to establish device-driver software. You might, for example, need to use a CD-ROM drive, in which case this must be notified as a device in a line that starts *DEVICE=* and continues with the location of a driver file. Here are some examples taken from one modern PC machine:

DEVICE=C:\DOS\HIMEM.SYS

— loads the driver for using extended memory;

DEVICE=C:\DOS\EMM386.EXE NOEMS

— loads a driver which can make use of some pieces of memory between the 640K limit and which can also be used if a program needs expanded memory (by omitting the NOEMS portion);

DEVICE=C:\DOS\SETVER.EXE

— loads a driver which allows programs that specify a DOS version number to run with a later version;

DEVICE=C:\PANA\CDMKE.SYS /SBP:220

— loads a driver for the Panasonic CD-ROM drive;

DEVICE=C:\DOS\INTERLNK.EXE /DRIVES:3 /LPT:3 /NOSCAN /NOPRINTER

— loads a driver for a simple network connection to another computer with three disk drives, using a parallel port connection through LPT3;

DEVICE=C:\DOS\DRVSPACE.SYS /MOVE

— loads a driver that compresses data before recording it on the disk and expands the data as it is retrieved. This has almost the same effect as doubling the disk size.

Finally, all modern versions of MS-DOS include a SHELL line which is used to notify the location of the COMMAND.COM file that contains many of the MS-DOS commands.

The AUTOEXEC.BAT file

The AUTOEXEC.BAT file is another file that sets up the computer, but in a way that tailors it for what you require, rather than for a specific country or piece of hardware, though the setting for the keyboard is done using a line in AUTOEXEC.BAT. This file exerts its control *after* the computer has loaded up the MS-DOS operating system and *after* it has run CONFIG.SYS. You can run AUTOEXEC.BAT at any time while the computer is running simply by typing AUTOEXEC (press ENTER). This is not always desirable, but it can be useful.

AUTOEXEC.BAT is a form of file called a *batch file*, all of which are written in plain text and which have filenames with the extension BAT. Such files consist of MS-DOS commands, and they can be run as if they were program files of the COM or EXE type, simply by typing the main name and pressing the ENTER key.

No two AUTOEXEC.BAT files will be identical, because both computers and computer-users have different requirements. What follows is a list of some typical lines that are likely to be found in the AUTOEXEC.BAT files of machines running MS-DOS 6.22. You are likely to alter your AUTOEXEC.BAT file many times in the lifetime of the computer because the AUTOEXEC.BAT file reflects the way in which you use programs on the machine. You are likely to want to alter your AUTOEXEC.BAT file each time you install a new program, so that your AUTOEXEC.BAT file grows longer as the months go by.

For example, the AUTOEXEC.BAT file on the machine I am using to type this work appears as:

```
@ECHO OFF
PATH C:\MSDOS;C:\;C:\windows;C:\BATS;C:\editor
set dircmd=/w/p/o:gn
set temp=C:\temp
LH /L:1,15904 C:\MSDOS\keyb uk,,C:\msdos\keyboard.sys
LH /L:1,6400 C:\MSDOS\doskey/insert
LH /L:1,640 keyclick
LH /L:1,56928 C:\msdos\mouse
LH /L:0;1,42400 /S C:\MSDOS\SMARTDRV.EXE 3072 512
lh C:\msdos\undelete /tc-20
win
```

and many of the commands in this set are peculiar to this particular machine and to the way that I work with it. In addition, some lines have been inserted by MS-DOS 6.22 Setup and others by the action of other programs.

Lines that start with LH have been modified by the use of MEMMAKER, a program that frees as much as possible of the base memory by loading short driver routines into extended memory. For example, the line that was originally typed as KEYCLICK has become *LH /L:1,640 keyclick*. This is part of the MEMMAKER action, and you do not need to create the LH part of such lines by typing them, nor do you need to understand the significance of the numbers.

For your own AUTOEXEC.BAT file the most important line is likely to be the *PATH* line. This determines where the machine will search for batch and program files after trying the current directory (the directory you are currently using). The form of the command is PATH followed by a file path and using a semicolon to separate this from the next filepath. For example:

PATH C:\;C:\MSDOS;C:\BATS

ensures that the C:\ root directory is searched first, then the C:\MSDOS directory and then the C:\BATS directory. You can extend this as much as you need, but remember that having a very long PATH line can slow down the action of the computer each time you type a program name without any specific path. With C:\MSDOS in the PATH line you will never again need to log onto the MSDOS directory to run the Editor, for example, because if you type EDIT (ENTER) the machine will look for this file and find it in the MSDOS directory.

Another line, mentioned earlier, that you are almost certain to need in your AUTOEXEC.BAT file is:

KEYB uk,850,c:\dos\keyboard.sys

which loads in a memory-resident program, KEYB, that configures the machine to use the UK keyboard (with its £ sign). Some computers make use of the US keyboard and do not need this line — if the line is used with a US keyboard you will find that some keys produce the wrong symbols.

This line assumes that the C:\DOS directory contains all of the necessary files, and since MS-DOS 6.22 creates this directory and fills it, the assumption is reasonable unless you are using an earlier version of MS-DOS. The KEYB command is a standard DOS command, and the UK portion specifies that the UK keyboard is to be used. The number 850 is a 'code page' number and, unless you are using equipment of IBM manufacture, the number is not of much significance. The last portion provides the name of the file, KEYBOARD.SYS and its path, and is essential unless KEYBOARD.SYS is placed in the root directory of the hard disk. If this command is wrongly used you will get an error message 'Bad or missing keyboard file' while the machine is loading up. If you do not see the message, incorrect action is indicated by inability to make the Shift-3 key of a UK keyboard produce the £ sign.

The most fundamental commands of MS-DOS are all contained in a file called COMMAND.COM, so that if this file is not on your hard disk or is corrupted you will be unable to use these commands. Other commands that can be found in the AUTOEXEC.BAT file are *Echo, Pause, Mode, Rem, Date, Time, Prompt, Set,* and *Subst*.

Echo is used either in the form *ECHO OFF* to suppress the printing of the commands of the AUTOEXEC.BAT file on the screen. It is nowadays used in the form *@ECHO OFF*, which suppresses even the ECHO line. The other form of echo allows messages to be delivered in the AUTOEXEC.BAT file, such as:

ECHO Press any key to continue

which would place on the screen the message that follows the command word ECHO.

PAUSE is used to cause AUTOEXEC.BAT or any other batch file to pause executing commands until you press a key, and is used if you need to read a message. This can be used if a choice has to be made whether to install a driver or not, or for testing a batch file.

MODE is seldom (or never) used nowadays. It is a multi-purpose command for altering settings relating to screen, printer, serial port and keyboard use. It has fallen into disuse because its actions are normally nowadays carried out by other programs.

MODE can set the screen output to 40 or 80 characters per line, and to colour or monochrome (assuming the screen *can* be switched). A parallel printer can be set to print 80 or 132 characters per line and 6 or 8 lines per inch, assuming that the printer is capable of such settings. A serial output can have its serial protocols of Baud rate, parity, number of data bits and number of stop bits set so as to suit a serial printer. Output to the parallel port can be re-directed to the serial port as a way of printer-switching. The letters that follow the MODE command decide which peripheral will be affected, and some typical codes which are used are listed below.

> LPT1: Parallel printer, port 1
> COM1: Serial output, port 1
> 40, 80, CO40 or CO80 : Screen character/colour options

For a printer, MODE will use the port name, usually LPT1, followed by the number of columns per line and the number of lines per inch, with the *P* option added to prevent time-out messages being printed if the printer is slow. For example: *MODE LPT1: 132,8* will set the parallel printer to 132 characters per line, 8 lines per inch. The default is 80 character per line, 6 lines per inch. To set only the number of lines per inch, omit the first figure but keep the comma.

For serial port control, MODE is followed by COM1 or COM2 (one of these may already be used by an internal modem); COM3 and COM4 if they are fitted can also be specified. The COM specifier will be followed by figures for:

Baud rate, using 110, 150, 300, 600, 1200, 2400, 4800, 9600 and 19200.
Parity, which is *N* (none), *O* (odd) or *E* (even).
Databits, the number of bits per word, 7 or 8.
Stopbits, the number of stop bits, 1 or 2
— and also the *P* specifier to prevent time-out signals

The default settings are even parity, 7 data bits and 1 stop bit (2 if 100 Baud is selected).

Example: *MODE COM1:1200,N,8,1* sets up to use the serial port with 1200 Baud, no parity, 8 data bits, 1 stop bit. This is a very common combination of settings for modem use.

The REM word means reminder, and anything following REM will be ignored when AUTOEXEC.BAT runs. This can be used for notes to remind yourself of the purpose of a line, or it can be used as a way of temporarily disabling a command. If you place REM ahead of the first word in a command line, the command will be taken as a comment and ignored. This can be a useful way of finding out if a problem is caused by a particular command.

Date and time commands are seldom used in AUTOEXEC.BAT nowadays; they were common in the machines of the early 1980s before hardware clock circuits were used. By placing a *DATE* line and a *TIME* line in AUTOEXEC.BAT you were prompted to enter the current date and time, which would be maintained while the computer was running. Since the introduction of the AT machines in 1982, a hardware clock circuit has been used with a battery backup to keep time data in CMOS RAM, so that the computer can read this information as it is booted up. The *TIME* command can be used each time the clocks are put forward or backward, but it is no longer used in AUTOEXEC.BAT.

Prompt can be used to alter the shape of the prompt, the screen message that signifies that a command is expected. The default for modern machines is C:\>, but by using the *PROMPT* command in AUTOEXEC.BAT you can add other information, such as a date, or the directory as well as the drive.

SET, used alone, will print on screen all of the existing *environmental settings*, meaning settings that provide information for programs. Usually these will include COMSPEC, the file that has to be used for commands (usually COMMAND.COM) and PATH, showing the PATH set in the AUTOEXEC.BAT file.

Example: *SET TMP=D:* means that a program which uses the word TMP to represent temporary storage space can make use of drive D:

Example: *SET PATH=C:\COMPILE* means that a program which uses PATH to define where it will look for programs will use C:\COMPILE rather than follow the paths that have been set up in a PATH command.

SUBST is a command that is never used nowadays and which can cause considerable problems with modern programs. It allows you to substitute a letter for a valid but unused drive in the place of a path. For example, *SUBST K=C:\FILES\BOOKS* would allow a save command to use K: to specify the BOOKS directory. The old commands SUBST and JOIN should be deleted from any machine that uses Windows.

Directory management

When hard disk drives first came into use, the amount of storage that they offered, though large compared to the capacity of a floppy disk at that time, was not enormous by modern standards (20–40Mbyte), but it soon became obvious that new methods of maintaining a program directory were needed. If we used the system that is applied to floppy disks the number of files on a hard drive would make it impossible to see all the titles on one screen display, and there would be no way of grouping files in the way that one floppy, for example, can hold one set of related files.

Directory *trees* are the MS-DOS method of making life tolerable for the user of the hard disk. The principle is to subdivide the directory system so that using the MS-DOS DIR command does not result in page after page of file listings and that saving a file does not cause problems of conflicting program names.

The system uses file groups, called *directories* (or *sub-directories*) which are named like files. Using the command word DIR then shows the directory (group) names (but not the names of the files in them), along with the names of files in the main (root) directory. You can then call for a directory of one directory group only, greatly reducing the effort that you need to spend on finding anything useful. One considerable advantage of keeping files in a directory structure is that the number of files that can be stored in a sub-directory is limited only by the disk capacity. When a large number of small files are placed in the root directory of a hard disk, there is a danger of exceeding the number of filenames that can be used long before the disk is full. By using a sub-directory this is avoided. In general, only a few essential files (such as CONFIG.SYS and AUTOEXEC.BAT) should be placed in the root directory.

A hard drive can be pictured as a directory tree, like a family tree, with the root directory as the ancestor of all the files, Figure 15.6.

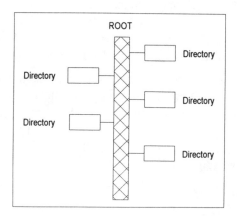

Figure 15.6   The root directory diagram, showing the sub-directories as branches from the root

However you like to imagine it, the diagram indicates that when you call for a directory display, all that you will get initially is the set of names that appear in the *root directory*, indicated as C:\ for a hard drive. Some of these names can be filenames of programs or data, and they will be indicated in the usual way. Other names can be of new directories, each containing its own files. These directory names are indicated by the <DIR> following the name of the directory in place of the usual extension of a filename. Another way of indicating these directory names is to enclose the name itself in angled or square brackets.

Only a few directories need to branch from the root. One directory can branch from another, so that one is the parent directory and the other is the child (Figure 15.7). A child directory is often called a sub-directory — in this sense, every directory apart from the root is a sub-directory, and it is possible for a sub-directory to be a parent to another sub-directory and a child of another.

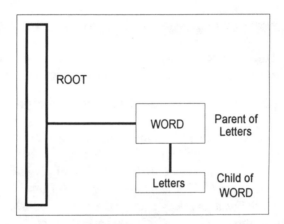

Figure 15.7   Illustrating parent and child directories. You can think of all other directories as being children of the root directory

The root directory, C:\, is the one to which you have immediate and automatic access when you start to use the hard disk. Unless you have some pressing reason to place program or data files in the root directory, you should always reserve the root for storing mainly sub-directories. In other words, when you are using the hard disk as drive C:, typing DIR should give you a list like that of Figure 15.8, in which each item other than a few files appears between square brackets, signifying a directory.

Because the amount of storage is large — it is not unusual to have 200–5,000 files on a hard disk — you must use some planning, deciding on how you want to group your programs. You might, for example, group programs according to their uses, such as word processors, spreadsheets, databases, graphics and so on, with perhaps another directory for combined programs such as MS Office. You should avoid placing data files in the root directory, and in many cases you would not

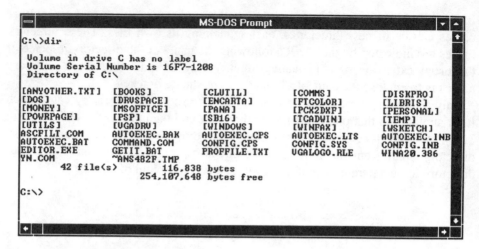

```
═                         MS-DOS Prompt                    ▼ ▲
                                                              ⬆
C:\>dir

 Volume in drive C has no label
 Volume Serial Number is 16F7-1208
 Directory of C:\

[ANYOTHER.TXT]   [BOOKS]        [CLUTIL]       [COMMS]       [DISKPRO]
[DOS]            [DRVSPACE]     [ENCARTA]      [FTCOLOR]     [LIBRIS]
[MONEY]          [MSOFFICE]     [PANA]         [PCX2DXF]     [PERSONAL]
[POWRPAGE]       [PSP]          [SB16]         [TCADWIN]     [TEMP]
[UTILS]          [VGADRV]       [WINDOWS]      [WINFAX]      [WSKETCH]
ASCFILT.COM      AUTOEXEC.BAK   AUTOEXEC.CPS   AUTOEXEC.LTS  AUTOEXEC.INB
AUTOEXEC.BAT     COMMAND.COM    CONFIG.CPS     CONFIG.SYS    CONFIG.INB
EDITOR.EXE       GETIT.BAT      PROFFILE.TXT   UGALOGO.RLE   WINA20.386
YN.COM           ~ANS482F.TMP
        42 file(s)         116,838 bytes
                       254,107,648 bytes free

C:\>
                                                              ⬇
◄ ◄▌                                                        ▌► ►
```

Figure 15.8   **How a DIR display appears when the root directory has been selected. The names in square brackets are directory names**

want to place data files in the same directories as the programs, but in sub-directories.

The root directory is indicated in commands by using the backslash sign, \. The root directory in the C: drive is therefore indicated by using C:\, and the root directory in the D: drive is indicated by D:\. The use of the backslash immediately following the drive letter (or immediately following a directory tree command) always means that the root directory is to be used. The backslash is also used as a separator to show the progression from a parent directory to a child, so that WORD\TEXT shows that TEXT is a child of WORD. Note that the letters A and B are reserved for floppy drives, with C onwards for hard drives, CD-ROM, network drives, and so on.

You have to be careful to distinguish these uses when you first start using a hard disk. The backslash symbol can be used by itself, so that DIR \ will produce a directory of the root directory. Take care over your use of the \ symbol, because careless use can lead to mistakes such as erasing the entire contents of the root directory. Note that Windows 95 refers to directories as *folders*, following the nomenclature used for the Apple Mac.

Directory maintenance

There are three commands that are concerned with directory use. CD (or CHDIR) is used to display or to change the directory you are using (the current directory). MD (or MKDIR) is used to create a new directory that branches off the directory you are using. RD (or RMDIR) is used to delete a directory that contains no files and which you do not want to use again.

CD used alone (or followed by drive letter only) will display the current directory path that is being used in that drive. CD followed by a drive letter can be used to show the current path in a drive that is not being used at present (there must be a disk in the drive). For example, if you are currently using the directory whose path is C:\WRITING\LETTERS, then typing CD (press ENTER) will produce on screen:

C:\WRITING\LETTERS

as a report. You can also use a line such as *CD A:* to show what path you are using in another drive (the A: floppy disk in this example).

CD is much more commonly used to change from one directory to another, to make another directory the current directory. When you use CD in this way, you must follow the command by a space and then the path to the directory you want to use. For example:

CD C:\WRITING\BOOK1

will take the path from the root hard drive to BOOK1, leaving this as your current directory.

Using MD to create a sub-directory applies also if you are starting from another sub-directory. If you switch to the WORD sub-directory, for example, you can create the further sub-directories of BOOKS, ARTICLES and LETTERS. Each of these will be a 'child' directory of the WORD directory.

You have to be in the WORD directory in order to create these sub-directories of WORD. If you are in the root directory when you use MD BOOKS, for example, you will certainly have created a sub-directory called BOOKS but it will be a sub-directory of the root, at the same level as WORD rather than as a sub-directory of WORD. This implies that if you are setting up a hard disk from scratch by transferring a large number of files from floppy disks it is a considerable advantage to plan carefully first, using a tree diagram of the type that has been illustrated. As an exercise you should draw the tree diagrams for these examples. Without such a diagram, it is remarkably easy to create sub-directories that you don't need, and put files in places where you can't easily use them.

You can also create a new directory on any other part of the hard disk or on a floppy. For example, using:

MD C:\WORD\LETTERS\NEWLET

will create a sub-directory called NEWLET if the directories of WORD and LETTERS already exist, and the LETTERS directory has no existing sub-directory called NEWLET.

The most difficult part of using a directory tree consists of closing down a directory that you no longer need. You might want to do this because you have found a better way to organise your directories, or because the programs in the directory have been superseded by others. There are, however, several important rules about removing a directory. To start with, you cannot delete the root directory in any circumstances, nor can you delete the sub-directory that you are currently using. You cannot delete by normal methods a directory that contains files, and you cannot delete by normal methods a directory that contains a sub-directory. There is a specialised command DELTREE (MS-DOS 6.0 onwards) that allows you to delete directories that contain files, but you should not use it until you have considerable experience.

It is always desirable to start any deletion operations from the root directory. If, taking the conventional example, you wanted to close a C:\WORD\BOOKS sub-directory that contained files with the TXT extension (but no further sub-directories), you would need to start this by using CD C:\ to get to the root directory. You would then use *DEL WORD\BOOKS*.TXT* to delete all of the files that use the TXT extension — note the use of a wildcard. You can then type *RD WORD\BOOKS* to delete the BOOKS sub-directory.

When you delete a set of files using a wildcard, you should always select the files by some common factor, such as the TXT extension in this example. If you use DEL *.* you will delete every file in a directory. This is fine if you really want to delete every file, but it is a disastrous command if you are currently using the root directory. You are always prompted to reconsider the DEL *.* command, and you should always do so.

You cannot delete the sub-directory until you have first deleted all the files in the sub-directory, using RD BOOKS, not DEL BOOKS. RD is short for RMDIR (remove directory), and it cannot, remember, be applied to the directory you are presently using. In other words, you cannot remove sub-directory BOOKS if you are working in this directory, you must work either from the root directory or from another sub-directory that is not a child (or any other descendent) of BOOKS, in this example.

TREE is the command that is used to display directories and sub-directories. To print the output your printer should be able to cope with the IBM character set — most dot-matrix printers can be arranged to emulate the IBM ProPrinter, and H-P Laserjet or compatibles can be set for the PC-8 character set.

If TREE is used alone it will display the tree diagram for the current drive and directory. Using TREE C:\ will usually result in a display that is too large to display on one sheet of paper or one screen, and you can use TREE C:\ | MORE to break the display into pages. You can also use TREE on part of a directory structure. For example, TREE C:\MSOFFICE will display only the MSOFFICE directory and branches from this directory.

Other fundamental commands

The PC machine, running MS-DOS, is ready for a command when its screen shows a 'prompt', usually consisting of the drive letter, followed by the colon, backslash, and > sign. Since the use of a hard disk drive C for booting the machine is almost universal, the normal MS-DOS prompt is $C:\backslash>$. When you use MS-DOS 6.22, the prompt will include the directory that is being used. With the C:> prompt showing, any standard MS-DOS command can be used, starting with the command name. You then need to type any supplementary information, the 'parameters' (additional information) and finally press the ENTER key to start the action.

For example, you can see on the screen the contents of a directory by typing:

DIR

and pressing the ENTER key. Since this command uses no parameters, the machine interprets it as meaning you want to see the directory listing of the directory that is currently in use. If the command is issued in the form:

DIR A:

the parameter A: means that the directory of the disk in the A: drive will be shown. Similarly, using

DIR C:\WORD

will display the contents of the WORDS directory in the C: drive. If you type only *DIR \WORD* this will be taken to mean the same.

Some commands are used without parameters, others need several parameters, and you need to know both the commands and what parameters can be used. Commands come in two forms, internal and external. The internal commands are built into MS-DOS and are available simply by typing the command and any parameters that are needed, irrespective of what drive or directory you are using. These internal commands are held in the COMMAND.COM file, part of which remains in the memory of the PC after booting. External commands make use of program files, one for each command.

When you issue any MS-DOS command that is *not* internal, you must specify in some way where the program file for that command is situated. For a computer with a hard disk, the default is to use whatever directory is currently in use, and the directories that are specified in the PATH line of AUTOEXEC.BAT. If the program file is in some other directory there are two ways in which this information can be specified. One is to specify it in the command itself. For example, if the main MS-DOS files are placed in a directory called C:\MSDOS then the external command FORMAT can be obtained by typing:

C:\MSDOS\FORMAT

305

File management commands

The most fundamental and the simplest file action is copying one disk file to another disk or directory. You can do this either retaining the name, or with renaming. The form of the command is:

COPY SOURCE DESTINATION

with the files *SOURCE* and D*ESTINATION* specified. The words must be separated by at least one space. Example:

COPY C:\WORDS\myfile.txt A:\

will copy a file called MYFILE.TXT in the C:\WORDS directory so that it is placed on a disk in the A: drive, using the same name of MYFILE.TXT.

There *must* be a space between the command and the source, and between the source and the destination. If you do not specify a destination filename, the current directory and drive is assumed; but this must not be where the source file is located — you cannot copy a file to its own drive and directory because this breaks the MS-DOS rule about having only one file with a given name. If only a drive letter or path is shown as the destination, the file is copied to this drive or directory using the same name.

You can use a *wildcard* in the name, and the first file to fit the description will be copied. A wildcard means the characters ? or *, with the ? meaning any single characters, and * meaning any set of characters. For example, if you use FILE.TX? this could refer to FILE.TXT, FILE.TXA, FILE.TXB and so on. Using FILE?.TXT could refer to FILE1.TXT, FILE2.TXT, FILE3.TXT and so on. Using NAME.* could refer to NAME.TXT, NAME.DOC, NAME.BAT and so on. For example, using:

COPY A:CHAP?.* C:\WPFILE

will copy files from the disk in the A: drive, using any file that begins with the letters CHAP with any single character (1, 2, 3 etc.) following, and with any extension letters. These files will be copied to the WPFILE directory on the hard drive.

The COPY command is the most-used of all MS-DOS commands because copying is one of the most frequently needed actions. It can copy data or program files to/from one disk or directory to another, text files to the screen, or text files to the printer. It can also alter the date and time of files and combine several text files into one single copy.

XCOPY is used for more specialised copying actions. For example, XCOPY can be used in a way that allows a file to be tested to find if it has been altered since it was last saved, and to save the file only if it has been changed. This allows XCOPY to be used as a way of automatically backing up data.

DEL is the command used to delete a file or a set of files. The DEL command is typed with a space separating this from the name of the file that you want to delete. DEL can be used along with wildcards, but you must be careful to avoid deleting files that you might have preferred to keep. Files can be undeleted if you act in time, because when a file is 'deleted' its directory entry on the disk is changed, but not the stored file itself. The file is deleted only if another file is stored in its place, overwriting the file. Un-deletion is possible only if nothing else has been saved.

You can use DEL in forms such as

DEL C:\TEXT\MYFILE.TXT

to delete this named file in this directory. This will not affect any file of the same name in another directory. Similarly, using:

DEL C:\FOLIOS\CHAP?.TXT

will delete files such as CHAP1.TXT, CHAP2.TXT, all the way to CHAP9.TXT, but not CHAP10.TXT or higher numbers. Once again, only files in this stated directory will be affected. You can also use the asterisk wildcard in the form:

DEL C:\NEWWORK*.DOC

to delete all files that use the extension of DOC and which are contained in this directory. This use of DEL is more risky, because if the directory contained a large number of files you might not notice a few that you have preferred to keep.
Going one step further, you can type:

DEL C:\ALLWORK*.*

to delete all of the files in this directory, regardless of the main name or extension. This is one of the most risky uses of DEL, and when you use this form you will be asked to confirm that you really want to delete all the files in the directory.

RENAME, usually abbreviated to REN, is used to rename a file, or a set of files. You might use this because you want the files, which currently have no extensions or which use different extensions, to carry a standard extension such as TXT. Another reason might be that you need to impose some common pattern on the names, such as changing ONE, TWO and THREE into CHAP1, CHAP2 and CHAP3.

Taking other examples, you might need to make use of a backup file called, for example, MYTEXT.BAK, but this cannot be read by your word-processor until it has been renamed MYTEXT.DOC, or you might need to alter a sequence, so that files called FIG1, FIG2, FIG3 should be renamed as FIG2, FIG3, FIG4

The REN command has the form:

REN oldname newname

meaning that you type the name REN, a space, the name of the file as it exists at the moment, another space and then the new name you want to use. In its simplest form this presents no problems, so that the command:

REN myfile.TXT oldfile.DOC

will change the name of MYFILE.TXT to OLDFILE.DOC. You will no longer find a directory entry for MYFILE.TXT

As you might expect, you cannot carry out a renaming if there is already a file in the same directory that carries the name you want to use. In the above example, there must not be a file already existing with the name OLDFILE.DOC. The filename that you want to use must be valid by the rules of MS-DOS.

ATTRIB is used to change file attributes. These are single-bit codes held in one byte of each file. The most important are the Read-only, Archive, System and Hidden attribute bits (often written as RASH). There is also one bit that determines whether a file name is used for a directory or for a file in a directory. The actions of these bits, when set to logic 1 are:

R attribute	Make file read-only
A attribute	The file has been altered since it was last backed-up
S attribute	The file is a System file, used by MS-DOS or Windows. System files are not normally listed in a DIR display.
H attribute	The file is a Hidden one which does not appear in a DIR display.

The Attrib command can be used either to find what attribute bits are set, or to change the attribute bits. For example:

ATTRIB C:\testfile

tests the file C:\testfile. You will see a line appear under the command when you press the RETURN key. If this is:

R C:\testfile

you know that the file is read-only.

The command:

ATTRIB +H C:\testfile

will make C:\testfile into a hidden file. The command ATTRIB -R C:\testfile will allow the file to be written as well as read. All of the attribute letters can be specified as + or − like this to turn an attribute on or off. Changing attributes is much easier if you use the Windows system.

BACKUP and RESTORE are commands that can be used for data security, but their use is not common. Early versions were unreliable, and though the most recent versions in MS-DOS 6.22 are much improved, the system uses floppy disks. Backing up a normal size of hard drive in this way needs 70 or more floppy disks, and is very time-consuming. Tape streamers using tape cartridges of 350 Mbyte or more are not expensive and are extensively used for backup purposes; a more expensive alternative is the read-write optical disk.

MS-DOS 6.22 no longer uses a separate RESTORE command, and for backing up or restoring data the MSBACKUP command is used. There is also a Windows version MWBACKUP. When either version runs you can select from a menu whether you want to make a full or partial backup, or to restore data from floppy disks.

For many purposes, a backup of data that has been changed (using the XCOPY command) is as much as is needed, providing that you have backup copies of programs and of older data files.

Integrity software

Older versions of MS-DOS use the CHKDSK command as a way of checking the integrity of hard or floppy disks. CHKDSK, used alone, will check the tracks of the current drive, and you can add a drive letter following the CHKDSK command to specify another drive or directory. CHKDSK must never be used if SUBST or APPEND commands have been used, which is why these commands are not supplied in modern copies of MS-DOS.

CHKDSK is used to report on how the storage space is allocated among directories, hidden files, normal files and free space, and it will also show, when it is applied to a file, how that file is split up for storage purposes. Optionally, CHKDSK will also attempt to mend disk faults that it finds, and will issue one or more of a large set of error messages if faults are found.

Example: *CHKDSK C:* produced the display for an old hard drive —

Volume FILECARD created 5 Feb 1988 14:14
32610304 bytes total disk space
110592 bytes in 6 hidden files
165888 bytes in 71 directories
20834304 bytes in 1492 user files
11499520 bytes available on disk
655360 bytes total memory
179680 bytes free

This shows the analysis of files and directories, and also the memory analysis, something that can be useful. The comparatively small amount of memory shown as free reflects the use of the machine — the CHKDSK program was run while a word-processing program was also being run.

A filename can be specified as well as a disk drive and path, and the report will then end with a note on the file. For example, a file report might contain the information that the file consisted of five non-contiguous (not touching) blocks.

This would be an indication that the file has become very scattered in the course of editing, and that access to it will be slower.

CHKDSK has two command line options. /V will report on progress as the disk is being checked and /F will repair damaged files as far as is possible. Using the /F option means that the CHKDSK process can be slow, as any damaged files will be repaired. Sections of files which cannot be allocated anywhere are placed into files with names such as *FILE1.CHK, FILE2.CHK* and so on. These are useful only if they are text files, because program files cannot be put together again and run.

CHKDSK is no longer supplied with MS-DOS and its use has been superseded by the later SCANDISK command, which checks the disk surface more thoroughly. Both CHKDSK and SCANDISK will find any defective pieces of disk and ensure that they are not used.

FORMAT prepares a floppy disk for use, or wipes a previously used disk clean. This is one of the few commands that (in its Unconditional option) can totally delete data beyond hope of recovery. It can be used for a hard disk, and though it will not necessarily destroy all of the data beyond recovery (depending on the disk partitioning), hard disk users should ensure that FORMAT is not present on the disk. FORMAT is needed only once on a hard drive, when the disk is put into service, and it must not be used again unless the disk is being stripped of all data.

FORMAT is therefore used on new blank floppy disks to prepare them for using in a PC machine. Note that a disk that has been formatted by another type of machine, such as an Apple Mac, will not necessarily be useable by a PC machine, though it is possible to prepare interchangeable disks.

As an example: FORMAT A: will format a floppy in the A: drive. After formatting, you will be asked if you want to format another disk. The modern version of FORMAT will check the disk structure and format it accordingly (in other words, for a 3.5" disk it will check if the disk is intended as a 720K or a 1.4M type of disk).

There are many options, several of which are of little interest except to users of odd disk formats. The most useful option is /S which puts system tracks on to the disk while formatting. This *must be the last option* if several options are specified. For example, FORMAT A: /S will format a new floppy in the A: drive and place the MS-DOS system and hidden files on it. This disk can be used to boot the computer if the hard drive fails, and several such 'system disks' should be made each time a new version of MS-DOS is installed.

FDISK is an obsolete command that is no longer included in MS-DOS. It was formerly used for the old types of hard drive that needed a 'low-level format', and it must on no account be used on modern IDE (integrated device electronics) drives. Considering the normal life of a hard drive, machines currently in service

310

are almost certain to use IDE or EIDE drives. FDISK was also used at one time to *partition* hard drives when the older MS-DOS versions could not cope with drive sizes more than 32Mbyte. Using FDISK allowed a 90Mbyte drive, for example, to be used as if it were three separate drives (letters C:, D: and E:) each of no more than 32Mbyte. Partitioning is no longer needed on drives of 2Gbyte or lower capacity.

DISKCOMP is intended for floppy disk users only, and should not be applied to a hard disk, nor to RAMdisk nor to networks. It allows two floppy disks to be compared, noting any differences. The command is by now obsolescent. DISKCOMP must not be used if ASSIGN, JOIN or SUBST have been previously used and not cancelled. When the program has run it presents on screen or as a file, the differences between the contents of the disks.

DISKCOPY is used for floppy disks to make a complete copy of one disk to another. The destination disk should preferably be formatted and unused, but if it contains files, these will be over-written. If it has not been formatted, then formatting will be carried out before copying — but this takes longer. The source disk from which files are to be copied should *always* be write-protected before starting to use DISKCOPY, particularly if a single floppy drive is being used. DISKCOPY has no application to a hard disk.

The important point about DISKCOPY is that it copies byte-by-byte with no regard for files, so that system tracks, hidden files, deleted files, corrupted files etc. will all be copied across to the destination disk. The disks must be fully compatible in layout — you cannot use DISKCOPY when one disk is a 3½" 720Kb type and the other is a 5¼" 1.2Mb type, for example. Always use DISKCOPY if you want to make an *exact* copy of a disk which contains files you want to work on with a disk utility program (if, for example, you want to 'undelete' files). It is also very useful for making exact backups of the disks on which software is distributed. When you use DISKCOPY with only one floppy drive, as is now usual, you are prompted to take out the disk you are copying and put in a blank formatted disk. Using modern MS-DOS versions, this is needed only once for each disk that is copied. Example:

DISKCOPY A:

uses single drive only, and prompts you to change between source and destination disks as required.

You can use, if you want, an unformatted disk for copying to, and it will be formatted in the process of making the copy. On modern machines it is not usually necessary to change between source and destination disk more than once.

Virtual disk

A virtual disk or *RAMdisk* means that part of the memory of the computer is used as if it were a disk drive. This allows for very fast reading and writing, but since the memory is always cleared when the machine is switched off, the files have to be loaded into RAMdrive from a real disk drive before use and restored to disk after alteration. On PC machines, the RAMDRIVE or VDISK drivers are loaded by using a DEVICE line in the CONFIG.SYS file, such as:

DEVICE=RAMDRIVE.SYS

RAMdrive is now an obsolete system because of the higher speeds of modern hard drives and the use of extended memory for multi-programming with Windows. In addition, the use of cache memory has superseded RAMdrives. Cache memory is used as an intermediate storage between disk and microprocessor, allowing large amounts of data to be held temporarily in memory, so that they can be read from or written to at high speed. The MS-DOS SMARTDRV driver is used to install a cache on modern computers.

Pipes and filters

The idea of piping and filtering was introduced to MS-DOS from other operating systems, notably Unix. A pipe is a way of moving data to a different destination, and MS-DOS uses the < and > symbols for pipe actions. For example, the DIR A command will provide a screen display of the root directory contents of the A: drive. If you want this information placed into a file called C:\DIRFILE, you can use the command:

DIR A:\ > C:\DIRFILE

which pipes the data that would normally appear on the screen into the specified file — note that the data will not then appear on the screen. You can also pipe to a printer using > PRN in place of > C:\DIRFILE in the above example. Piping out, as illustrated here, is much more common than piping in using the < symbol, because very few programs are written to allow piping in.

The filter action makes use of a program called a filter, meaning one that alters data and allows piping both into the program and out of it. One example of a simple filter is called MORE, and it can be used in the form:

TYPE filename | MORE

where filename represents the full filename (with path) of a text file. The effect of this change is to display the file on the screen in pages, so that you can read a page and then press any key to see the next page. When the filter is used between

312

the command (TYPE in this example) and its normal output (the screen in this case), the vertical bar is used to indicate filtering.

You can also use the < and > signs to place a filter between a command and an output. For example, the filter SORT will, as the name suggests, sort a list of words into alphabetical order. The words are typed, one word (or phrase) on each line, and saved as a file. If we give this file the name UNSORT, then the command:

$$SORT < UNSORT > SORTED$$

will take the unsorted file, sort it and place the result into a file called SORTED. You can pick for yourself the names you want to use for the files.

NOTE: For a comprehensive treatment of MS-DOS commands from version 3.3 to 6.22, see *Newnes MS-DOS Pocket Book* from Butterworth-Heinemann. For information on Windows 95, see the *Pocket Book of Windows 95* from the same publisher.

Test Questions

1. State what is meant by a GUI system of operation. Give two reasons for preferring this to the older type of DOS.
2. State what is meant by the base memory of a PC machine. What name is used for the majority of memory in a machine with 8Mbyte of RAM?
3. A program is loaded into a PC that uses the 8086 processor, and a utility finds that 150Kbyte of RAM is available for data. How is the rest of the base memory used?
4. How much storage space would you expect to allocate for a screen of 640 x 480 pixels that displays in 16 colours per pixel? Show your working.
5. State how you could deal with the following mouse problems:
   (a)   the mouse cursor sticks or moves jerkily,
   (b)   no mouse cursor appears and moving the mouse has no effect.
6. Describe the command sequence that is implemented when you type a DOS command and press the ENTER/RETURN key.
7. State two important differences between the AUTOEXEC.BAT and the CONFIG.SYS files.
8. You have just installed a tape streamer to back up your hard disk. What type of command would you expect to be added to the CONFIG.SYS file?
9. What is the purpose of a PATH line in the AUTOEXEC.BAT file?
10. In the directory whose tree is drawn in Figure 15.9, your current directory is the C:\ root. Write the command line that would make the ARTICLE directory current.

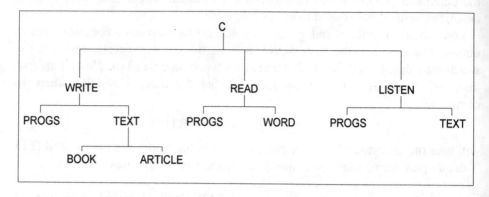

Figure 15.9   Diagram for question 10

Appendix 1   Programs in BASIC

Program 1

```
10 REM 80x24 SCREEN OF Xs
20 CLS
30 FOR n = 1 TO 1920
40    PRINT "X";
50    NEXT n
60 END
```

Program 2

This provides Red, Green, Blue or White rasters which are useful for purity adjustments and general signal tracing.

```
10 REM SCREENS OF RED, BLUE, GREEN OR WHITE
20 CLS
30 COLOR 1,7
40 INPUT "ENTER R B G or W for colours. Q to stop";A$
50 IF A$="R" OR A$="r" THEN F%=4
60 IF A$="B" OR A$="b" THEN F%=1
70 IF A$="G" OR A$="g" THEN F%=2
```

```
80 IF A$="W" OR A$="w" THEN F%=7
90 IF A$="Q" OR A$="q" THEN 160
100 B%=F%
110 COLOR F%,B%
120 FOR N=0 TO 45
130   PRINT "                              "
140 NEXT N
150 GOTO 20
160 CLS
170 END
```

Note that in line 130 there should be 40 spaces between the quotes.

Program 3

```
10 CLS; REM GRID & CIRCLE for CGA & EGA
20 GOSUB 1000;CLS
30 IF a$="Q" OR a$="q" THEN END
40 SCREEN M;PSET(1,1)
50 FOR a=0 TO 639 STEP 40
60    FOR B=0 TO Ht
70          PSET(a,B)
80      NEXT B
90 NEXT a
100 FOR B=0 TO Ht
110   PSET(a-1,B)
120 NEXT B
130 FOR B=0 TO Ht STEP S
140   FOR a=0 TO 639
150         PSET(a,B)
160   NEXT a
170 NEXT B
180 FOR a=0 TO 639
190   PSET(a,Ht)
200 NEXT a
210 Pi#=4*ATN(1!)
220 CIRCLE (320,C),R
230 LOCATE 12,35;PRINT"PRESS any KEY"
240 IF INKEY$="" THEN GOTO 240
250 SCREEN 2
```

```
260 GOTO 10
1000 REM select modes
1010 PRINT "Do you want CGA(15.5kHz)=C"
1015 PRINT" or EGA(21.6kHz)=E";PRINT; PRINT"Q to QUIT"
1020 PRINT; PRINT"Enter the letter";PRINT
1030 a$=INPUT$(1)
1040 IFa$=""GOTO 1030
1050 IF a$="C" OR a$="c" THEN
        M=2;Ht=199;S=20;C=100;R=200;RETURN
1060 IF a$="E" OR a$="e" THEN
        M=9;Ht=349;S=35;C=175;R=200;RETURN
1070 IF a$="Q" OR a$="q" THEN SCREEN 2;RETURN
1080 CLS; GOTO 1010
```

Note: in lines 1050 and 1060 the line should be typed as one, so that the portion starting M= is part of the main line. It has been broken up on this page so that the lines do not run off the end of the page.

Appendix 2   Answers to test questions

Answers to Test Questions for Chapter 1

1. Fan-out = 1.6/0.2 = 8.
2. Fan-out value depends on clock speed because the current is required to drive stray capacitances.
3. ECL is the most suitable technology for this clock speed.
4. CMOS is appropriate because of its low-power operation. Speed is not important.
5. Noise threshold is important, and the use of 15V supplies provides a larger threshold.
6. Open-collector gates must be used.
7. This is a CMOS chip, and you would expect $0.6\mu W$ dissipation per gate and propagation time of around 40–250ns.
8. The range +2V to +5V is acceptable.
9. (a) SSI, (b) VLSI, (c) VLSI.
10. 4001 CMOS, 74LS240 Low-power Schottky, 74HC10 High-speed CMOS, 7400 Standard TTL.

Answers to Test Questions for Chapter 2

1.   BCD, 0110,1001 = 69. Pure binary 0110,1001 = 1 + 8 + 32 + 64 = 105.
2.   1000,1111 = – (1 + 2 + 4 + 8) = – 6 + 15 0101,0000 = +(16 + 64) = + 80.
3.   By continued division 576(10) = 10,0100,0000. Arranged into groups of 4 bits and adding leading 0s, this becomes 0010,0100,0000 = 240 in both Hex and BCD.
4.   Complement of 10A3H = EF5C+1 = EF5D, Adding, F002 + EF5D = 1DF5F and after dropping the leading 1 overflow, this becomes DF5F.

Answers to Test Questions for Chapter 3

1.   See Figure A1, below.

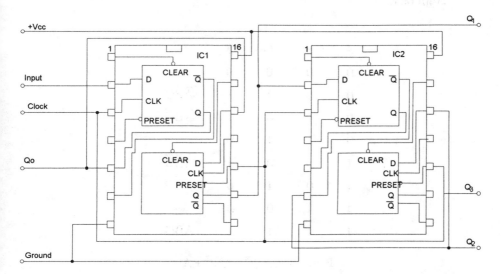

Figure A1   The connections between the chips for answer to question 3.1

2. This question is related to odd parity checking.
   The truth table is:

A	B	C	X	Y
0	0	0	0	0
0	0	1	0	1
0	1	0	1	1
0	1	1	1	0
1	0	0	1	1
1	0	1	1	0
1	1	0	0	0
1	1	1	0	1

3. This question is related to majority voting control.
   Truth table:

A	B	C	F	
0	0	0	0	
0	0	1	1	$\overline{A}\overline{B}C$
0	1	0	1	$\overline{A}B\overline{C}$
0	1	1	0	
1	0	0	1	$A\overline{B}\overline{C}$
1	0	1	0	
1	1	0	0	
1	1	1	1	ABC

$$F = \overline{A}\overline{B}C + \overline{A}B\overline{C} + A\overline{B}\overline{C} + ABC$$

From this the Karnaugh map can be derived:

AB \ C	00	01	11	10
0	0	0	1	0
1	0	1	1	1

$F = AB + AC + BC$

The logic diagram is shown in Figure A2.

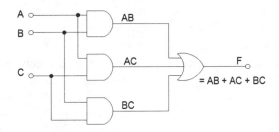

Figure A2   The logic diagram for question 3.3

4.   This device is very similar to that shown in Figure 3.11 except that it uses NAND gates as the output elements.
Truth table:

A	B	Y0	Y1	Y2	Y3
0	0	0	1	1	1
0	1	1	0	1	1
1	0	1	1	0	1
1	1	1	1	1	0

A binary input pattern ranging from 00 to 11 will uniquely select 1 of 4 output lines, Y0 to Y3. The output lines will be active Low and can be used as chip select lines (CS) for 4 EPROMs covering the addresses 0000H to 0FFFH, 1000H to 1FFFH, 2000H to 2FFFH and 3000H to 3FFFH.

Answers to Test Questions for Chapter 4

1.   ROM must be present to provide the first instructions to the processor, and RAM must be present to store other instructions and data that are read in from a disk.
2.   A clock circuit using CMOS RAM has been running powered by a battery to maintain the time.
3.   Possible if the reprogramming involves only the severing of more connections. A severed connection cannot be re-joined.
4.   EPROM can readily be re-programmed. Changes in a masked ROM require re-tooling and are very expensive.

5.   Dynamic RAM stores charge in tiny MOS capacitors. These leak, requiring the refresh action. Static RAM is costly and power-consuming; the low-power CMOS type is too slow for most computer uses.

6.   Use the same address connections for both PROMs and store the upper bytes in one PROM and the lower bytes in the other.

7.   Storage capacity for each chip is given from the number of address lines, which is 11. This is 2^{11}, equal to 2,048, and since the three chips are used independently, the total number of bits is $3 \times 2,048 = 6,144$ bits.
      The ROM addresses start at 0000H and end at 07FFH, the EPROM range is 0800H to 0FFFH, and the RAM range is 1000H to 17FFH

8.   EPROM can be erased by using ultra-violet light, EAROM can be erased by using higher supply voltage and cycling through the addresses.

9.   An EPROM such as the 2732 could be used, held in a programmer that allowed data for each address to be typed on a number pad, using a key to increment the address number after each byte of data. Another method is to use a computer interface and type the data as prompted by the computer software.

10.   Dynamic memory is comparatively slow and needs refresh cycling at 1ms intervals. By using a different bank for each byte in a set, the computer can maintain its rate while still allowing the memory to recover and be refreshed. This is aided by using a cache which is filled at a steady rate and can be used to feed bytes to the microprocessor at a very fast rate.

Answers to Test Questions for Chapter 5

1.   (a)   The resolution is determined by the number of bits per sample.
      (b)   Conversion rate = $\frac{1}{2} \times 10^{-6} = 500$ks/s.
            Highest frequency component = $500 \times 10^3/8 = 62.5$ kHz.
            frequency will be reduced to 31.25kHz, showing how a trade-off can be managed.
      (c)   This converter requires pulses from 1 to 2^{n-1} to convert from 1 bit to full scale, i.e. 1 to 255 pulses.
            Conversion clock period = 10^{-4}s = 100μs. Therefore time ranges from 100μs to 25.5ms.

2.   (a)   Nyquist sampling frequency = 2×8kHz = 16kHz.
      (b)   In practice, the sampling frequency would typically be increased to 20kHz.
      (c)   0.1% = 1/1000. The nearest convenient value would be 1/1024.
            $1024 = 2^{10}$, 10 bits therefore give a resolution of 0.0977%.

(d)   See Figure A3. The aperture time represents the time taken for the sample switch to operate after the start of the pulse. In 100ns, the voltage rises at 2V/µs or 200mV/100ns. Hence this represents a signal level loss.

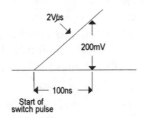

Figure A3   Illustrating aperture time

3.   (a)   The code 1000,0000 represents half full scale. Therefore the dynamic range is given by Maximum volts/Resolution 10V/39mV = 256.4 or 20log 256.4 = 48.18 dB.

(b)   Number of levels = 2^8 = 256.

(c)   Each level oF the resolution = 10V/256 = 39mV.

(d)   Output bit rate = 8 x 1Khz = 8kbit/s.

(e)   Maximum theoretical conversion rate = 500Hz.

4.   (a)   See answer 2(c) earlier.

(b)   See answer 2(c) earlier.

(c)   Nyquist clock rate = 2 x 10kHz = 20kHz.

(d)   Output bit rate = 10 x 20kHz = 200kbit/s.

(e)   Minimum theoretical bandwidth = 200/2 = 100kHz.

5.   (a)   See Figure A4.

(b)   See Figure A4.

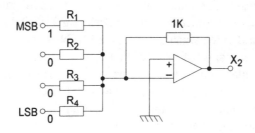

Figure A4   Implementation of D/A converter circuit

(c)   See Figure A5. Op-amp gain = $-\frac{1}{4}$, therefore feed back resistor = 8k/4 = 2k.

6. (a) See Figure A4. If Op amp gain is $x2$ then MSB resistor = 500Ω. Therefore R1 = 500Ω, R2 = 1k, R3 = 2k and R4 = 4k.
    (b) 1000 represents half full scale, therefore full scale value is 8V. 4 bits provide 16 levels, therefore the resolution is $8/16 = 0.5V$.
    (c) The important feature of a D/A converter is accuracy. The binary weighted resistors values range from relatively low to significantly high. For a 10-bit converter, this can range from 10k to over 5M. Over this wide range, it is very difficult to achieve a balanced temperature tracking. If the highest value is reduced to a low enough manageable level, then the low value becomes comparable with the switch output resistance to compound the error further. The choice of values is therefore a compromise of temperature dependence against accuracy.

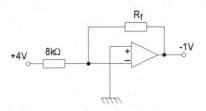

Figure A5   Values for answer 5 (c)

Answers to Test Questions for Chapter 6

1. The ISDN system operates entirely in the digital mode whilst the PSTN system is essentially analogue based. The ISDN signals are carried over coaxial cables rather than twin parallel or twisted pair cables. Thus the signals are better protected against interference. ISDN readily provides for data, voice and video signals, whilst the PSTN is very bandwidth restricted, which limits the range of services that can be carried.
    From the digital point of view, the bit rates over ISDN is significantly higher. Due to the use of addressed and numbered data packets, the several packets of a complete message may be transmitted over several different paths to avoid noisy parts of a network.
2. DTMF is more computer compatible. Allows different key combinations to control a remote computer system. Provides faster dialling. Allows the use of on-hook dialling.
3. ASK signals are bandwidth restrictive and have noise-like characteristics which makes them more difficult to separate from noise. PSK is a minimum

bandwidth system that allows the transmission rate to be increased. FSK is commonly used for magnetic tape and disk modulation.

4. Logic 1 is represented by 8 cycles of 2.4kHz and logic 0 by 4 cycles of 1.2kHz. Both frequencies conveniently fit within the audible bandwidth of the PSTN. The bit periods are of equal time duration. This is a form of FSK with better noise immunity characteristics than ASK.

Answers to Test Questions for Chapter 7

1. (a) 16 colours requires 4 bits per pixel and there are 640 x 480 pixels per screen. Total number of bits = 640 x 480 x 4 = 1,228,800 bits or 153,600 bytes.
   (b) Each character requires two bytes, one for foreground and one for background. Therefore total number of bytes = 80 x 25 x 2 = 4,000 bytes.
   (c) Number of pages = 38.4.
2. (a) Primaries: Red, Green and Blue. Secondaries: Magenta, Cyan and Yellow.
   (b) With the loss of Green, the colour bars change as follows: White, Yellow, Cyan, Green, Magenta, Red, Blue, Black.
   Magenta, Red, Blue, Black, Magenta, Red, Blue, Black.
3. Wear safety goggles and gloves. Monitor should be switched off, disconnected from power supply and all the associated capacitances should be discharged. Tube should not be handled by neck. The new tube should be retained in its case until needed and the displaced one should be suitably packed as soon as it has been removed.
4. Provides a support for the CRT and a common earth point. Restrains the atmospheric forces on the domed tube face — hence it is also a safety feature.
5. Using the Nyquist sampling rate, the bit rate for each channel is: 2 x 8MHz x 8 = 128Mbit/s. For the three channels this becomes 128 x 3 = 384Mbit/sec and the bandwidth = 192MHz.

Answers to Test Questions for Chapter 8

1. (a) A display of data, usually in hexadecimal for each line of a bus.
   (b) A display of the waveforms on each line of the bus. Hex displays can be on an LCD panel or on a CRT, waveform displays are usually on a CRT.

2. See Figure A6
3. (a) The probe tip is placed on a bus line to indicate logic levels of 0, 1, or pulsing.
   (b) The pulser connector is placed on a bus line to inject pulses so that their effect can be checked.
4. So that a changing voltage can be followed.
5. It can display only one or two traces, operates at relatively low speeds, and has no storage facilities.
6. Count cycles of both the clock and the input wave until a coincidence occurs for a whole number of cycles of each.
7. A buffer to prevent loading the clock circuits (hence causing mistiming) by the load of the other circuits.
8. Such stages will provide 'clean' pulses from inputs that have unacceptably long rise and fall times.
9. This is the result of a check-sum calculation on all the ROM contents, and can be checked by repeating the check-sum calculation.
10. The emulator provided inputs that might be difficult to provide under working conditions, and it allows operation will all mechanical parts disabled.

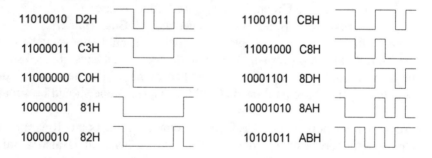

Figure A6   Hex equivalents and waveforms

Answers to Test Questions for Chapter 9

1. Periodic time = 2 x 3 x 33ns = 198ns. Frequency = 1/Periodic time = 1/198ms = 5.0505 MHz. See Figure A7 for diagrams and waveforms.

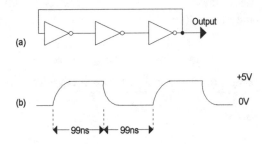

Figure A7   Diagram and waveform for answer 9.1

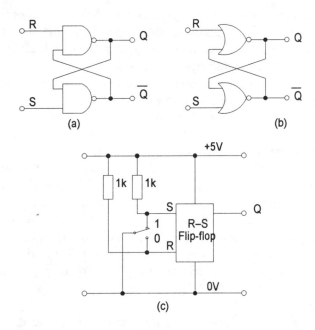

Figure A8   Switch debouncing circuits for answer 9.2

2.   See Figure A8.
3.   (a)   A mechanical flip switch is included within the casing of the floppy disk. The position of this either exposes or closes an aperture, a state which is sensed by an optical coupler to inhibit or enable the ability to record.
      (b)   See Figure A9. The ID address pattern is used to identify the beginning and so synchronise the serial data processing action. The

error check may be either a simple check sum or a full cyclic redundancy check (CRC).

I.D. Address	Track Number	Side Number	Sector Number	Sector Length	CRC

Sector

I.D. Address	Data (128 – 256 – 512)	CRC

Figure A9   Address pattern drawings for answer 9.3(b)

    (c)   The process of formatting a disk lays down the necessary track and sector magnetic patterns for future use. Thus when being loaded with actual data, the synchronism addresses and data locations can readily be found by the disk operating system.

    (d)   Track 00 provides the reference point for data synchronism.

4.   (a)   Printers use a range of different control character sequences to select different print modes, formats and fonts. For ease of connection to a range of computers, each printer is often provided with the facilities to emulate one of the more popular devices.

    (b)   Dot matrix printers in particular, commonly have at least two print quality modes, draft and NLQ. Draft mode printing is carried out in a single pass with the print head traversing the paper from left to right and then right to left on the next line of text (Bidirectional). This operation considerably increases the printing speed. In the NLQ mode, printing is a two pass operation with the head passing twice in the same direction (Unidirectional) but with the paper slightly shifted to minimise the print dottiness.

    (c)   See Figure A10.

    (d)   The important features include; control of several disks, automatic verification of disk, formatting ability, automatic sector search, complete track read or write, read or write a number of blocks.

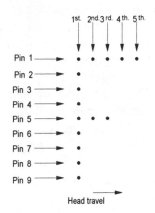

Figure A10   Pin pattern for answer 9.4(c)

Answers to Test Questions for Chapter 10

1. Cluster is a term used to describe a terminal configuration of several devices connected to a single node of a network. For example, this may be a work-station consisting of a personal computer, printer, plotter, scanner, etc. All of these may be capable of working together independent of the network, or working through the network with other users.

2. (a) Advantages: networking makes better use of expensive peripheral devices as these can be shared with other users. Since data base information can be shared with other users, networking improves the dissemination of data throughout an organisation. Wide and Global area networking allows communications over greater distances and reduces personal travelling.

    Disadvantages: initial installation costs can be high. However, these can usually be recovered through the many advantages. It is important to ensure that the system is user friendly otherwise the network is not fully utilised. The major problem is associated with hacking. This can provide the opportunity for malicious damage to the network security, access to secret files, falsification and corruption of the data base through viruses, theft of money or corporate information.

    (b) Protocols represent a set of internationally or widely agreed rules by which different devices or systems are allowed to communicate with each other. These rules set the way in which the data is to be

329

represented; organised for transmission; transmission data rate controlled; timed and synchronised to avoid unwanted interaction with other users; addressed to the appropriate user; data errors introduced by noise and distortion controlled.

3. (a) Amplitude shift keying is represented by discrete changes of signal levels and is closely related to amplitude modulation. Other features being equal, ASK occupies the greatest bandwidth.

Frequency shift keying involves shifting the carrier signal between discrete values. In the binary case, just two frequencies are employed and the bandwidth is simply the difference between the two.

Phase shift keying employs a single carrier frequency whose instantaneous phase is shifted between specific values. The bandwidth is thus the narrowest of these three.

Quadrature amplitude modulation involves a combination of amplitude and phase modulation. Each specific allowable signal phase can also take up a number of unique amplitudes.

(b) QAM allows several bits to be transmitted for each transmission symbol, and thus increases the overall data rate without increasing the bandwidth.

(c) A baseband system transmits the raw data and thus occupies a relatively low frequency spectrum. In broadband systems, the raw data is modulated on to a carrier and produces a wider bandwidth signal. This, however, allows many signals to be frequency multiplexed to maximise the use of the transmission medium.

4. (a) The earthed outer sheath of the coaxial cable screens the signal carrying inner conductor from any interference signals. These signals thus flow to earth over the single outer conductor.

With twin cables, the wanted signals flow alternately through both wires equally. Any interference signals flow to earth through both wires simultaneously, but in the same direction. The electromagnetic effect that this creates is thus self cancelling.

(b) Serial transmission need only a pair of conductors and is thus simple. By comparison, parallel transmission requires a path for each bit, thus increasing the cabling costs, but providing a much higher data rate. The simpler serial mode in general, allows for a greater transmission path length, but at a lower data rate.

(c) The two logic level thresholds range from 5 to 15 volts; positive for logic 0 and negative for logic 1. The noise and attenuation margin thus lies between +5 and −5 volts.

Answers to Test Questions for Chapter 11

1. The 2-byte operand could be an address for an 8-bit system
2. A bus in which the data type alters at each strobe pulse, for example, address bits on one pulse and data bits on the next.
3. The use of chip-select pulses to enable or disable the chips.
4. (a) to allow the system to shut down in the event of a fatal error (such as a power fault).
   (b) to allow a sub-routine to be run with a correct return to the conditions at the time of the interrupt.
5. A polled system interrogates the keyboard whether a key has been pressed or not; an interrupt-driven system responds only when a key is pressed.
6. Direct Memory Access — this operates independently of the microprocessor so avoiding the need to use the microprocessor to read and write each unit of data. Also known as cycle-stealing.
7. The bus lines must be allowed time to settle to their correct logic levels.
8. Chip selection and address decoding
9. An expansion bus may need to accept cards that can run only at lower clock rates. In addition, the higher stray capacities of an expansion bus make it undesirable to run such buses at high clock rates.
10. (a) MPU, (b) Address decoder, (c) ROM.

Answers to Test Questions for Chapter 12

1. Static RAM is preferred on grounds of speed. A cache built into a microprocessor is usually small, typically 16Kbyte, so that the power requirements are not so much of a problem as they would be for a main RAM of 8Mbyte or more.
2. Memory-oriented MPUs can use a small number of 8-bit registers, with memory used in place of larger registers. Memory access is very fast, with a specified area of memory reserved for use as a set of registers. Register-oriented MPUs possess a large number of registers, most of which can operate either as single-byte registers or in combination for larger units.
3. Only CMOS MPUs can be operated at low speeds. Other types must be operated at high speeds because of the leakage between tracks on PMOS or NMOS devices.
4. A sequence of programs steps for an important action, such as ADD, that is triggered by an instruction code. Using a microprogram system avoids the need to have long and complex instruction codes, and ensures that only the correct codes will affect the MPU.

5. An instruction queue is a set of instructions codes cached inside the MPU; a cache can contain both instruction codes and data and can be internal or external.
6. One input of the ALU is always from the accumulator, and the output of the ALU (following an ALU action) is taken to the accumulator.
7. These bits can be tested by MPU instructions, and the results used to trigger other actions, such as a jump to another address.
8. See Figure 12.5 in the main text.
9. The segment address is shifted left by one hex digit, making it A0000, and the offset is added to give A1FC0 as the final address.
10. This allows programs that were written for the older processors (using a total of 1Mbyte of RAM) to be run, one in each 1Mbyte of this partitioned space.

Answers to Test Questions for Chapter 13

1. (a) To allow efficient programming when a piece of code, such as a time delay, needs to be run more than once in the course of a program.
   (b) To allow the design of programs to be split into sections for easier planning and writing.
2. See Figure A11

Figure A11   Flowchart diagram for answer 13.2

3.   See Figure A12

Address	Data	HL	0800	0900
0400	21	xxxx	AF	xx
0401	00	xx00	AF	xx
0402	08	0800	AF	xx
0403	22	0800	AF	xx
0404	00	0800	AF	xx
0405	09	0800	AF	AF

Figure A12   Table of bus contents for answer 13.3

4.   LDA #FFH
Loop: DEC
     BNE Loop
     RTS
Delay: the instructions use $2 + 255 \times 6 + 255 \times 2 + 5 = 2{,}047$ cycles. At a clock rate of 1MHz (1µs pulses) this takes 2.047 ms.

5.   The register layout and the instruction format of the 8086 and the Z80 are very similar.

6.   Register-oriented processors possess a large number of registers on the chip, whereas memory-oriented types use RAM addresses as registers. The 80486, like all modern designs, is register-oriented.

7.   (a) 0D 08   (b) F7 D0 05 01   (c) D1 E8

8.   See Figure A13 below.

9.   See Figure A14 opposite.

10.   Save the current Program Counter address number, so that a return is possible.

PC	A	HL	ADDR1	ADDR2	ADDR3	ADDR4
7E7D	xx	xxxx	3F	xx	6B	xx
7E7E	94	xxxx	3F	xx	6B	xx
7E7F	7E	7E94	3F	xx	6B	xx
7E80	7E	7E94	3F	xx	6B	xx
7E81	83	7E94	3F	xx	6B	xx
7E82	41	7E94	3F	3F	6B	xx
7E83	xx	7E94	3F	3F	6B	xx
7E84	01	7E01	3F	3F	6B	xx
7E85	7F	3F01	3F	3F	6B	xx
7E86	7F	3F01	3F	3F	6B	xx
7E87	80	3F80	3F	3F	6B	xx
7E88	41	4180	3F	3F	6B	6B

Figure A13   Trace table for Answer 13.8

Figure A14 Flowchart for Answer 13.9

Answers to Test Questions for Chapter 14

1. (a) Real-time use may require high-speed because the processor will need to be able to work at the highest rate imposed by the input(s). Off-line processing can be done at lower speeds.

    (b) On-line use often demands a large memory capacity, so that data can be loaded into the memory (using DMA) while the processor is working.

    (c) Off-line use demands large disk space, because all the stored information for a day's work may need to be stored for processing overnight.

2. A dedicated microcomputer can use only as much memory and other resources as are appropriate to the task. This is often provided on the MPU chip, so that the dedicated microcomputer consists of very little more than this main chip. A general-purpose microcomputer needs a large memory and hard disk size to cope with programs for which each new version is larger. The dedicated microcomputer will need inputs and output to serve the machine that it controls; the general-purpose computer often uses only one parallel port for the printer and a serial port for a modem, along with a mouse port (see Chapter 15).

3. Outputs will be to water valve, to water heater relay, drain pump, wash motor and spin motor. Inputs are from water level sensor and temperature sensor. The microprocessor will output logic 1 to the water input valve until the input from the water level sensor is logic 1. The heater will then be switched on until the output from the temperature sensor is logic 1. The wash motor is then switched on for a preset time (time-delay loop), then off. The drain valve is opened, and the spin motor started until another preset time has elapsed, when both are switched off. The water inlet valve is opened again until the level is correct, and the wash motor used again for a preset time (rinsing). The drain valve and spin motor are then used again, and this sequence repeated for as often as required. Finally, all output and inputs are disabled at the end of the cycle.

4. Connect an output port to a D/A converter circuit. Output to the port a number derived from a counter. For example, you could increment the number each 10μs, allowing a count of 9999, and at the final count figure reset the counter so as to return the output voltage at the port to zero.

5. The microcontroller chip will use a PROM to hold its program, and this chip must be connected to a PROM-blower circuit, for which address numbers and data are provided by the main computer.

Answers to Test Questions for Chapter 15

1. A GUI system shows, on the screen, icons (images) in place of menu items, and an action is activated by placing a pointer over an icon, using the mouse, and clicking the mouse button. Items (images or text) can also be moved by dragging, clicking over the item and holding the mouse button down while the mouse is moved. The advantages are that there is no need to type the name of a program in order to run the program (so that correct spelling is no longer necessary!), and that items can easily be selected from lists. The advantages are particularly important for graphics programs, but text-based programs also benefit. Users find it easier to learn to use a new program because the basic steps of using the mouse are common to all programs.

2. The base memory of a PC machine is the first 640Kbyte of RAM. This is used for running DOS programs. Later programs that use the Windows system (particularly since Windows 95) can use the whole of the memory which for a modern PC machine is usually 8Mbyte or more.

3. The rest of the base memory contains the program itself, the operating system, some memory-resident programs (keyboard, printer, mouse and screen drivers, for example), and buffer space.

4 Sixteen colours require 4 bits per pixel, so that one byte could handle two pixels. Since there are 640 x 480 pixels, the number of bytes is 640 x 480 /2 which is 153,600 bytes.

5. (a) Clean the mouse ball and rollers.
   (b) Check that the mouse driver software is the correct version and that it has been run.

6. The use of the command word locates a routine in DOS. This is run, and makes use of IO.SYS and MSDOS.SYS. These files in turn locate and activate routines in the BIOS. The codes in the BIOS are passed to the microprocessor for execution.

7. The AUTOEXEC.BAT file is a batch file that can be run only when MS-DOS has been loaded. The CONFIG.SYS file runs before MS-DOS loads. The AUTOEXEC.BAT file can be run at any time by typing its name and pressing the RETURN key. The CONFIG.SYS file can be run only when the computer is booted up.

8. A DEVICE= line would be added to describe the type of streamer and any setup parameters.

9. To indicate where MS-DOS is to look for programs when no specific path has been included in the name of the program.

10. CD \WRITE\TEXT\ARTICLE

Index